Introductory Mathematics for the Life Sciences

David Phoenix

Department of Applied Biology
University of Central Lancashire
Preston, UK

Taylor & Francis
Publishers since 1798

UK   Taylor & Francis Ltd., 1 Gunpowder Square, London EC4A 3DE.
USA Taylor & Francis Inc., 1900 Frost Road, Suite 101, Bristol, PA 19007.

British Library Cataloguing in Publication Data

A catalogue record for this book is available from the British Library
ISBN 0-7484-0428-7

Library of Congress Cataloging Publication Data are available

Cover design by Jim Wilkie

Typeset in Melior by Keyword Publishing Services Ltd

Printed in Great Britain by T.J. International, Padstow, UK

Contents

General Preface to the Series

The curriculum for higher education now presents most degree programmes as a collection of discrete packages or modules. The modules stand alone but, as a set, comprise a general programme of study. Usually around half of the modules taken by the undergraduate are compulsory and count as a core curriculum for the final degree. The arrangement has the advantage of flexibility. The range of options over and above the core curriculum allows the student to choose the best programme for his or her future.

Usually, the subject of the core curriculum, for example biochemistry, has a general textbook that covers the material at length. Smaller specialist volumes deal in depth with particular topics, for example photosynthesis or muscle contraction. The optional subjects in a modular system, however, are too many for the student to buy the general textbook for each and the small in-depth titles generally do not cover sufficient material. The new series *Modules in Life Sciences* provides a selection of texts which can be used at the undergraduate level for subjects optional to the main programme of study. Each volume aims to cover the material at a depth suitable to the year of undergraduate study with an amount appropriate to a module, usually around one-quarter of the undergraduate year. The life sciences was chosen as the general subject area since it is here, more than most, that individual topics proliferate. For example, a student of biochemistry may take optional modules in physiology, microbiology, medical pathology and even mathematics.

Suggestions for new modules and comments on the present volume will always be welcomed and should be addressed to the series editor.

John Wrigglesworth, Series Editor
King's College, London

Preface

Students are entering A-level and undergraduate life science courses with only GCSE mathematics. Many students do not possess a thorough understanding of the basic mathematical principles which are required in these courses and those that do understand the mathematics often have difficulty applying the principles to biological problems. These deficiencies are difficult to correct and can involve the need for intensive tutorial-based courses, but with increasing student numbers and decreasing staff time the support for material which lies 'outside' the standard life science curriculum is limited. This leads to many students struggling with basic concepts, such as concentration, and if courses include areas with a strong mathematical orientation such as kinetics, energetics or even pH calculations students tend to gain little, since their time is spent struggling with the mathematics; thus they often miss the biological importance of the material.

This book has been written after discussion with undergraduates to find out the areas with which they want help. It is intended to introduce essential mathematical ideas from first principles but without the use of mathematical proofs. In the body of each chapter are worked examples so that readers can apply the mathematics and develop their confidence. At the end of each chapter are a number of questions taken from biology and these allow students to try to apply the mathematics they have learnt. The emphasis is on essential mathematics, i.e. that which students will need at some time in most courses and some of which will be applied on a daily basis. Once the mathematics has been learnt, students need to apply it. It is useful to perform the following steps when facing a numerical problem:

(a) look at the problem and write down all the information that you have;

(b) write down what it is you want to know;

(c) work out what information is actually required and what is superfluous;

(d)   establish the link between what is wanted and what is known;

(c)   apply the mathematics and find the answer!

David Phoenix
Department of Applied Biology
University of Central Lancashire

1 Numbers

1.1 Introduction

Scientists must be able to take quantitative measurements and look for correlations within their experimental data. A scientist should therefore be able to manipulate numbers and have an appreciation of their relevance. The objectives of this chapter are:

(a) to introduce real numbers;
(b) to develop rules for the manipulation of numbers.

1.2 Real numbers

Real numbers may be represented by their position on a **number line** (Figure 1.1). All the numbers which lie on this line are termed **real numbers** and the set is represented by the symbol $\mathbb{R}$. Whole numbers (**integers**) are represented by the symbol $\mathbb{Z}$ and can be sub-grouped into positive ($\mathbb{Z}+$) or negative ($\mathbb{Z}-$) integers.

Negative numbers are written to the left of zero. The further a number is to the right, the bigger it is, so for exam-

On the number line, the further the number is to the right the bigger it is

Figure 1.1

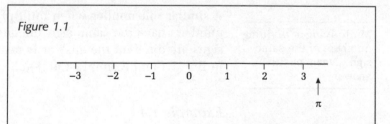

$\mathbb{R}$ represents the group of all numerical values which can be represented on the number line (i.e. the real numbers)
$\mathbb{Z}$ represents the set of intergers $\{\ldots -3, -2, -1, 0, 1, 2, 3, \ldots\}$
$\mathbb{Z}+$ represents the set of positive integers, sometimes called natural numbers ($\mathbb{N}$) $\{1, 2, 3, 4, \ldots\}$
$\mathbb{Z}-$ represents the negative integers $\{-1, -2, -3, -4, \ldots\}$

ple -2 is greater than -5. Addition therefore indicates that you move to the right, since the number is getting bigger; subtraction indicates that you move to the left.

It is obviously important that you are able to manipulate both positive *and* negative numbers. It is useful to remember that if you are adding a negative number to a positive number you can treat this as a subtraction, as shown in Example 1.1.

Example 1.1

$$(-2) + 3$$
$$= 3 - 2 = 1$$

It may help to remember the number line. In Example 1.1 you start at position minus two (-2) and plus three ($+3$) tells you to move to the right three places, which takes you to position one on the number line. In Example 1.2 you start at position minus four and move one place to the left, thus giving the answer minus five.

Example 1.2

$$-4 - 1 = -5$$

If you subtract a negative number it becomes positive

When dealing with negative numbers, the only rule that must be remembered is that if you subtract a negative number it becomes positive. This can be seen in Example 1.3.

Example 1.3

$$1 - (-3)$$ *Subtraction of a negative gives a*
$$= 1 + 3 = 4$$ *positive*

Multiplying or dividing numbers of the same sign gives a positive answer

A similar rule applies when multiplying or dividing; if both numbers have the same sign the answer is positive, if their signs are different the answer is negative. This is illustrated in Box 1.1 and Example 1.4(a)–(c).

Example 1.4

(a) $3 \times 2 = 6$ *Both signs are the same, therefore the answer is positive*

(b) $3 \times (-2) = -6$ *The signs are different, therefore the answer is negative*

(c) $(-9) \div (-3) = 3$ *Both signs are the same, therefore the answer is positive*

Box 1.1 **Sign rules for multiplication and division.**

(positive) × (positive) = positive	(positive) ÷ (positive)	= positive
(positive) × (negative) = negative	(positive) ÷ (negative)	= negative
(negative) × (positive) = negative	(negative) ÷ (positive)	= negative
(negative) × (negative) = positive	(negative) ÷ (negative)	= positive

If you have more than two terms in the calculation, then to apply the sign rules in Box 1.1 you need to break the calculation down into parts as shown in Example 1.5.

Example 1.5

$2 \times (-3) \times (-1)$ $2 \times -3 = -6$: *The different signs imply that the answer is negative*

$(-6) \times (-1)$ $-6 \times -1 = 6$: *The same signs imply that the the answer is positive*

$= 6$

Worked examples 1.1

Evaluate:
(i) 2×-5 (ii) -6×-3 (iii) $3 - 5$ (iv) $-2 - 6$
(v) $-3 - (-4)$ (vi) $-6 \div -6$ (vii) $6 \div -12$.

1.3 Modulus

On some occasions it may be the size of the value that is important, rather than its sign. For example, suppose you are measuring the height of a seedling in centimetres. The exact height is 4.7 cm and you take two measurements which are recorded in Table 1.1 along with the error.

Table 1.1

Reading (cm)	Error (cm)
4.5	−0.2
4.7	0.2

With the first reading you have under-estimated the height by 0.2 cm but the second reading is too large by 0.2 cm. The error in both cases is of the same size or **magnitude**; it is only the direction that is different, i.e. one is an under-estimate and the other an over-estimate. In this case it may be worthwhile considering the **absolute values**. The absolute value takes into account the magnitude or size of the change but

Modulus measures the
absolute value without
the sign

it does not take into account the direction of the change. It is denoted by two straight lines (i.e. $|-2| = 2$) and is usually called the **modulus**. In the example given above you can say that the **absolute error** in both measurements is 0.2 cm.

Worked examples 1.2

Evaluate:
(i) $-2 - |-2|$ (ii) $|3-5|$ (iii) $1 - 4 - |3|$
(iv) $3 + |2-3|$

1.4 Functions with multiple operations

You often have to deal with functions which contain more than one mathematical operation and it is important to know in what order to perform these operations. In general, if an expression contains brackets you always evaluate whatever is in the brackets first, then you perform multiplication and division and finally addition and subtraction (Box 1.2).

Box 1.2 **Priority of operations.**

> 1 Brackets
> 2 Multiplication and division
> 3 Addition and subtraction

If there is more than one set of brackets you start on the inside and work outwards.

Example 1.6

$((3-2) \times 4 + 4) \div 2$ *Innermost brackets first, so* $3 - 2 = 1$
$= (1 \times 4 + 4) \div 2$ *Brackets; multiplication, so* $1 \times 4 = 4$
$= (4 + 4) \div 2$ *Brackets; addition* $4 + 4 = 8$
$= 8 \div 2$
$= 4$

It is essential that these rules are applied since failure to do so will greatly influence the outcome of the calculation, as can be seen in the following examples.

Example 1.7

$3 + 4 \times 5 = 3 + 20$ *Compared with* $(3 + 4) \times 5 = 7 \times 5$
$= 23$ $= 35$

Example 1.8

$$6 - 4 \div 2 = 6 - 2 \qquad Compared\ with \qquad (6 - 4) \div 2 = 2 \div 2$$
$$= 4 \qquad\qquad\qquad\qquad\qquad = 1$$

Note that in Example 1.8 the expressions can be rewritten to emphasise their difference:

$$6 - 4 \div 2 = 6 - \frac{4}{2} \quad and \quad (6 - 4) \div 2 = \frac{6 - 4}{2}$$

In general, although the list of priorities tells you which operation to perform first, it is always best to use brackets to clarify what is required.

Example 1.9

$$6 - 4 \div 2 = 6 - (4 \div 2) = 6 - \frac{4}{2}$$

In Example 1.9 the brackets are not needed but their presence can help prevent confusion and this decreases the chance of error.

Worked examples 1.3

Evaluate:

(i) $3 - 9 \div 3$ (ii) $4 \times (2 - 3)$ (iii) $((4 + 6) \div 5 + 3) \times 3$
(iv) $10 \times 5 + 4 \times 5$ (v) $((15 - 5) + 2 \times 2) \div 7$.

1.5 Commutative and associative laws of addition and multiplication

The **commutative law** (Box 1.3) states that:

> *The order in which two numbers are added or multiplied may be interchanged.*

Box 1.3 **Commutative laws.**

$$a + b = b + a$$
$$ab = ba$$

If this law holds then the order in which we add or multiply two numbers does not matter since the order can be interchanged. Examples 1.10 and 1.11 show this to be true.

Example 1.10

$$2 + 5 = 7$$
$$5 + 2 = 7$$

Example 1.11

$$3 \times 2 = 6$$
$$2 \times 3 = 6$$

This law can be expanded to give the **associative law**. The associative law states:

If more than two numbers are added or multiplied it does not matter in which order they are added or multiplied.

Box 1.4 **Laws of association**

$$(a + b) + c = a + (b + c)$$
$$(ab)c = a(bc)$$

If, therefore, an expression contains only multiplication or only addition, the order in which the operations are performed is irrelevant. If you have been asked to evaluate this type of expression you can rearrange the calculation so that it can be performed in the easiest way possible, as shown in Example 1.12.

Example 1.12

$2 \times 17 \times 5$	*Compared with*	$2 \times 17 \times 5$
$= (2 \times 5) \times 17$		$= (2 \times 17) \times 5$
$= 10 \times 17$		$= 34 \times 5$
$= 170$		$= 170$

Both methods in Example 1.12 give the same answer but for most people the first route would be the easier one to follow. These rules also apply to subtraction and division, since these are simply inverses (i.e. the opposite) of multiplication and addition (Box 1.5).

Box 1.5 **Laws of association.**

$$a - b = (-b) + a$$
$$a - (b - c) = (a - b) + c$$
$$a \times \frac{1}{b} = \frac{1}{b} \times a \qquad (b \neq 0)$$

The order in which you perform multiplication and division does not matter if these are the only operations present

A consequence of the equations shown in Box 1.5 can be seen in Example 1.13. If the expression contains a mixture of multiplication and division, the operations can be separated and the order interchanged in the same way as in Example 1.12.

Example 1.13

$$\frac{17}{6} \times 6 = 17 \times \frac{1}{6} \times 6 \quad compared\ with \quad \frac{17}{6} \times 6 = 2.83 \times 6$$
$$= 17 \times 1 \qquad\qquad\qquad\qquad\quad = 17$$
$$= 17$$

The same applies to addition and subtraction, as is illustrated in Example 1.14.

Example 1.14

The order in which you perform additions and subtractions does not matter if these are the only operations present

$$3\left(-2\right) + 6\left(-4\right)$$
$$= 3 + (-2) + 6 + (-4)$$
$$= 3 + 6 + (-2) + (-4) \quad Rearrange\ employing\ the$$
$$\qquad\qquad\qquad\qquad\qquad law\ of\ association$$
$$= 3 + 6 + (-6)$$
$$= 3$$

Worked examples 1.4

Evaluate:

(i) $18 \times \dfrac{32}{9}$ (ii) $55 \div 13 \times 26$ (iii) $\dfrac{(16 + 17)}{11} \div 6$

Summary

Real numbers are values which can be represented by a point on the number line and the set of real numbers is described by the symbol $\mathbb{R}$. Integers are a sub-group of $\mathbb{R}$ and can be represented by the symbol $\mathbb{Z}$. Integers may be positive or negative but in some cases it is the magnitude of the value that is required and not its sign, and this is denoted by the modulus.

When performing calculations with multiple operations you always perform the calculation inside the brackets first, followed by multiplication and division and finally addition and subtraction. Since multiplication obeys the law of association, calculations containing only multiplication can be performed in any order and should be evaluated by the simplest route possible. The same rule applies to functions containing only addition.

End of unit questions

1. Calculate the following:
    (a) $20 \times 18.5 \times 5$
    (b) $0.6 \times 12.5 \times 5 \times 8$
    (c) $32 \times 5 \div 8$

2. Evaluate the following:
    (a) $4 - 7$
    (b) $-3 - (-2)$
    (c) $9 + 23 - 47 - 2$

3. If $a \times b = ab$ define the following:
    (a) $a \times -b$
    (b) $a \times -b \times -c$
    (c) $-c \times -b$

4. Calculate the following:
    (a) $(6 - 2) \div 4 + 7$
    (b) $22 \times 7 \div 11 + 6 - 3$
    (c) $(((24 - 14) - 5 \times 6) - 5) + 25 - 40 \div 8$

5. In an experiment on CO_2 evolution students were required to estimate the surface area of a leaf. The actual area was $16.3 \, cm^2$. The students' estimates were 10, 16, 19 and $23 \, cm^2$. Calculate the error and the absolute error in each case.

6. Ostwald's Dilution Law can be used to find the ionisation constant for weak electrolytes such as propionic acid.
    (a) Evaluate the calculation:

    $$\text{Ionisation constant for propionic acid} = \frac{0.037 \times 0.037 \times 0.01}{1 - 0.037}$$

    (b) Rewrite the calculation in one line, using brackets where necessary.

7. A Warburg manometer flask can be used to measure pressure changes when a gas is produced or used. For example, the uptake of oxygen by a bacterial suspension can be measured as the bacteria respire. Before this can be done the manometer constant for oxygen needs to be calculated for the experiment.

    $$\text{Manometer constant (oxygen)} = \frac{20\,000 \times \frac{273}{310} + 3000 \times 0.024}{10\,000}$$

    (a) Calculate the constant from the above equation.
    (b) Rewrite the expression on a single line using brackets to help clarify the operations.

2 Fractions, Percentages and Ratios

2.1 Introduction

Science rarely produces answers in the form of integer values so students must be able to break numbers down into parts or fractions and to have an appreciation of what a fraction represents. In addition you should be able to perform numerical operations with fractions such as addition, subtraction, multiplication and division. Fractions can be represented by decimals, and students should be able to interconvert decimals and fractions when the need arises. If work is being performed in which various compounds are combined (for example, a number of solutions could be mixed to provide the correct environment for a biological assay), then you should realise what fraction of the whole each component represents and be able to express this in the form of ratios and percentages. Since a variety of experimental data can be expressed as a percentage, it is important that fractions, ratios and percentages can be interconverted. The objectives of this chapter are as follows:

(a) to develop confidence in handling fractions, percentages and ratios;

(b) to develop an appreciation of their relationship to data;

(c) to be able to interconvert the three forms of expression.

2.2 Fractions—rational and irrational numbers

Fractions are represented in the form:

$$\frac{p}{q} \text{ where } \frac{p}{q} = p \div q$$

p and q are integers. q is called the **denominator** and p is termed the **numerator**. p is usually less than q so that the numerical value is less than one. This is called a **proper fraction** and an example is given (Example 2.1).

Example 2.1

$$\frac{2}{5} = 0.4$$

Any value which can be obtained by dividing two integers in this manner is called a **rational number** and is represented by the symbol $\mathbb{Q}$. All integers are therefore rational numbers, as shown in Example 2.2; but some values cannot be represented in the form p/q, for example the values of pi (π) and $\sqrt{2}$, and these are termed **irrational numbers**.

Example 2.2

$$\frac{3}{1} = 3$$

If the value of the fraction is greater than one, as in Example 2.2, then it is termed an **improper fraction** (Example 2.3).

Example 2.3

$$\frac{11}{4} = 11 \div 4 = 2 \quad \textit{with 3 remainder}$$
$$= 2\tfrac{3}{4} \quad \textit{or 2.75}$$

$\frac{11}{4}$ is therefore an improper fraction but it can also be expressed as the **mixed fraction** $2\frac{3}{4}$. A mixed fraction contains an integer value plus a proper fraction.

It is worth noting that:

$$\frac{0}{q} = 0 \quad \textit{but} \quad \frac{p}{0} \quad \textit{is meaningless}$$

Division by zero is not possible

The reason why q cannot equal zero is that division by zero is not defined. Since the denominator never equals zero, rational numbers are usually represented by the following expression:

$$\frac{p}{q} \quad (q \neq 0)$$

This is summarised in Box 2.1

Box 2.1

(i) $\dfrac{p}{q}$ represents a rational number

(ii) If p is less than q then it is a proper fraction

(iii) If p is greater than q then it is an improper fraction

(iv) $\dfrac{0}{q} = 0$ but q can never equal zero.

2.3 Factorisation and equivalent fractions

There are many cases in which you need to factorise an expression, i.e. write it as a product. In Example 2.4 the numbers nine and six have been factorised. It can be seen that both numbers can be written as a product which contains the factor three. Three is therefore said to be a **common factor** with respect to six and nine. Fractions can be simplified if both the numerator and the denominator have a common factor, and the factorisation method shown in this section can be used to find any common factors.

Example 2.4

$$9 = 3 \times 3$$
$$6 = 2 \times 3$$

This method uses **prime numbers**, i.e. numbers which are divisible only by themselves and one {2, 3, 5, 7 etc.}. In example 2.4 the integer nine has been written as a product of two prime numbers and is said to have been **prime-factorised**. To prime-factorise a number, try dividing it by the prime number two, and if the number is not divisible by this amount, try the next prime number in the series – three – and so on. For example, fifty is divisible by two but the other factor formed (twenty-five) is not a prime number; hence we need to repeat the process, as seen in Example 2.5.

Example 2.5

$$50 = 2 \times 25$$
$$= 2 \times 5 \times 5$$

You are now at the stage where all the numbers in the expression are prime numbers so when 50 has been prime-factorised it is represented as {2 × 5 × 5}.

A second instance is given in Example 2.6.

Example 2.6

$$35 = 5 \times 7$$

In this case neither two nor three divides into thirty-five, so the first prime number of use is five. Since seven is also a prime number, thirty-five has now been prime-factorised. Once you have a list of the prime factors of a number, you can find all its factors, i.e. all the values by which it can be divided. This set simply includes one, the number itself, the prime factors and all the possible multiples of the prime factors. This method of prime factorisation is used in Example 2.7

Example 2.7

$$18 = 2 \times 9$$
$$= 2 \times 3 \times 3$$

Factors of 18 *are therefore* {1 *and* 18} *plus the prime factors* {2 *and* 3} *plus multiples of the prime factors:*

$$2 \times 3 = 6 \text{ and } 3 \times 3 = 9$$

Hence all the factors of 18 *are* {1, 2, 3, 6, 9, 18}

The ability to use prime factorisation is especially useful when dealing with fractions. Example 2.8 shows how prime factorisation can be used to simplify large unwieldy fractions.

Example 2.8

$$18 = 2 \times 3 \times 3 \quad and \quad 24 = 2 \times 2 \times 2 \times 3$$

so

$$\frac{18}{24} = \frac{2 \times 3 \times 3}{2 \times 2 \times 2 \times 3}$$

employing the law of association

$$= \frac{3 \times (2 \times 3)}{2 \times 2 \times (2 \times 3)}$$

Common factors in the denominator and numerator can be cancelled

---

Both the numerator and denominator contain the common factors two and three; hence these factors can be cancelled:

$$\frac{18}{24} = \frac{3 \times (2 \times 3)}{2 \times 2 \times (2 \times 3)}$$
$$= \frac{3}{2 \times 2} = \frac{3}{4}$$

You can confirm that the above is true since:

$$\frac{18}{24} = 18 \div 24 = 0.75 \quad and \quad \frac{3}{4} = 3 \div 4 = 0.75$$

Since $\frac{18}{24}$ and $\frac{3}{4}$ are numerically equivalent they are called **equivalent fractions**. $\frac{3}{4}$ cannot be simplified further since there are no more factors common to the numerator and denominator. $\frac{3}{4}$ is therefore said to be in its **simplest form**.

It is worth mentioning that multiplying both the numerator and denominator by the same constant always gives an equivalent fraction (Example 2.9).

Example 2.9

$$\frac{18}{24} = \frac{3}{4} \quad as \quad 18 = 6 \times 3 \quad and \quad 24 = 6 \times 4$$

$$so \quad \frac{18}{24} = \frac{6 \times 3}{6 \times 4}$$

In Example 2.9 both numerator and denominator were multiplied by the same constant. This is not the same as multiplying the whole fraction by a constant, which would only increase the size of either the top or the bottom of the fraction and change the value, as is shown in Example 2.10.

Example 2.10

$$\tfrac{3}{4} \neq (6 \times \tfrac{3}{4}) \quad as \quad 6 \times \tfrac{3}{4} = 6 \times 3 \div 4$$
$$= 4.5$$

$$or \quad 6 \times \frac{3}{4} = \frac{18}{4}$$
$$= 4\tfrac{1}{2}$$

Multiplication of the numerator and denominator by a constant should not be confused with the addition of a constant, because even if you add the same constant to the top and bottom of a fraction the numerical value changes. This is illustrated in Example 2.11 and the results are summarised in Box 2.2.

Example 2.11

$$\frac{6+3}{6+4} = \frac{9}{10} \quad and \quad \frac{9}{10} \neq \frac{3}{4}$$

Box 2.2

$$\frac{a}{b} = \frac{Ka}{Kb} \neq K\left(\frac{a}{b}\right)$$

$$\frac{a}{b} \neq \frac{(a+K)}{(b+K)} \quad where \ K \neq 0$$

Worked examples 2.1

Simplify the following where possible:

(i) $\frac{9}{36}$ (ii) $\frac{27}{18}$ (iii) $\frac{24}{16}$ (iv) $\frac{3}{7}$

2.4 Addition and subtraction of fractions

For addition and subtraction of fractions you will need to find the **lowest common multiple** of two numbers, i.e. the smallest value into which both numbers will divide. In this case prime factorisation (Section 2.3) can be used to help. The method for finding the lowest common multiple is shown in Example 2.12 for twenty and eighteen.

Example 2.12

$$20 = 2 \times 2 \times 5 \quad and \quad 18 = 2 \times 3 \times 3$$

Two and five occur most often in the prime factorisation of twenty, which has as its factors two twos and one five. Three occurs most often in the factorisation of eighteen. There are no other factors present apart from these three. Let the lowest common multiple therefore contain two twos and one five from twenty, and two threes from eighteen.

$$2 \times 2 \times 3 \times 3 \times 5 = 4 \times 3 \times 15$$
$$= 3 \times 60$$
$$= 180$$

Hence 180 is the smallest number that is divisible by both twenty and eighteen.

Worked examples 2.2

Find the lowest common multiple of:
(i) 14 and 24   (ii) 18 and 33   (iii) 27, 18 and 54
(iv) 24, 18 and 33

If the operation of addition or subtraction is to be performed, then all fractions must have the same denominator, so the first step is to find the lowest common multiple for the denominators.

Example 2.13

$$\frac{1}{6} + \frac{3}{8}$$

Prime factorisation of the denominators gives:
$$6 = 2 \times 3 \quad and \quad 8 = 2 \times 2 \times 2$$

The lowest common multiple is therefore
$$2 \times 2 \times 2 \times 3 = 24$$

To convert $\frac{1}{6}$ into a fraction with a denominator of twenty-four, the denominator must be multiplied by four; hence the numerator must also be multiplied by four (Box 2.2).

$$\frac{1}{6} = \frac{1 \times 4}{6 \times 4} = \frac{4}{24}$$

The same procedure can be applied to $\frac{3}{8}$:

$$\frac{3}{8} = \frac{3 \times 3}{8 \times 3} = \frac{9}{24}$$

The fractions can then be added:

$$\frac{1}{6} + \frac{3}{8} = \frac{4}{24} + \frac{9}{24} = \frac{4+9}{24} = \frac{13}{24}$$

The same procedure is applied in the case of subtraction, as in Example 2.14.

Example 2.14

$$\frac{1}{6} - \frac{3}{8} = \frac{4}{24} - \frac{9}{24} = \frac{4-9}{24} = \frac{-5}{24}$$

2.5 Multiplication of fractions

This operation can be performed simply by multiplying the denominators and the numerators (Example 2.15).

Example 2.15

$$\frac{2}{11} \times \frac{1}{4} = \frac{2 \times 1}{11 \times 4} = \frac{2}{44} = \frac{1}{22}$$

It is worth noting that in this example the calculation could have been simplified since the numerator and denominator contain a common factor of two which can cancel.

$$\frac{2}{11} \times \frac{1}{4} = \frac{1 \times (2)}{11 \times [2 \times (2)]} = \frac{1}{11 \times 2}$$

2.6 Division of fractions

If you wish to divide one fraction by another, simply invert the dividing fraction and multiply them.

Example 2.16

$$\frac{9}{11} \div \frac{1}{3} = \frac{9}{11} \times \frac{3}{1} \quad \text{\textit{Inverting the dividing fraction}}$$
and multiplying

$$= \frac{9 \times 3}{11 \times 1}$$

$$= \frac{27}{11} \quad \text{\textit{or the mixed fraction }} 2\frac{5}{11}$$

Example 2.17

$$\frac{9}{11} \div \frac{3}{4} = \frac{9}{11} \times \frac{4}{3} \quad \text{\textit{Inverting and multiplying}}$$

$$= \frac{9 \times 4}{11 \times 3}$$

$$= \frac{(3 \times 3) \times 4}{11 \times 3} \quad \text{\textit{Note that} 9 \textit{and} 3 \textit{have the}}$$
common factor 3, *which cancels*

$$= \frac{3 \times 4}{11 \times 1}$$

$$= \frac{12}{11} \quad \text{or} \quad 1.09$$

Worked examples 2.3

Evaluate:

(i) $\frac{1}{3} + \frac{7}{8}$ (ii) $\frac{1}{2} - \frac{4}{10}$ (iii) $\frac{5}{7} - \frac{10}{12}$ (iv) $\frac{3}{4} \times \frac{2}{7}$ (v) $\frac{4}{11} \times \frac{22}{30}$
(vi) $\frac{6}{13} \div \frac{1}{2}$ (vii) $\frac{2}{3} \div \frac{1}{9}$.

2.7 Percentages

To convert a fraction or decimal to a percentage, multiply it by 100

A percentage represents a fraction of 100, i.e. a fraction with a denominator of 100. To convert a fraction to a percentage all you need to do is multiply it by 100. If the fraction is represented by a decimal, the same rule applies.

Example 2.18

$\frac{3}{4}$ *of a solution is used – what percentage of the total is this?*

$$\frac{3}{4} \times 100\% = \frac{3}{4} \times \frac{100}{1} = \frac{300}{4} = 75\%$$

To calculate a percentage, the first step is therefore to represent the value you require as a fraction or decimal.

Example 2.19

A DNA fragment of 35 kilobases is digested by an exo-nuclease. The enzyme degrades seven kilobases. What percentage of the DNA is degraded?

7 out of 35 kilobases are degraded, i.e. $\frac{7}{35}$, so the percentage is:

$$\frac{7}{35} \times \frac{100}{1}\% = \frac{700}{35} = 20\%$$

Note that in Example 2.19 the fraction $\frac{7}{35}$ can be represented by the equivalent fraction $\frac{1}{5}$ since both the denominator and numerator have the common factor 7. Using this equivalent fraction would have simplified the calculation:

$$\frac{1}{5} \times \frac{100}{1}\% = \frac{100}{5} = 20\%$$

Suppose instead that you wish to find a percentage of a given amount, for example 15% of 70. In this case convert the percentage to a fraction or a decimal and multiply it by the amount concerned.

Example 2.20

What is 15% of 70?

$$15\% = \frac{15}{100} \text{ so } 15\% \times 70 = \frac{15}{100} \times \frac{70}{1} \quad \text{100 } and \text{ 70 } have \text{ the} \atop common \text{ factor } 10$$

$$= \frac{15}{10} \times \frac{7}{1} \quad \text{15 } and \text{ 10 } have \text{ the} \atop common \text{ factor } 5$$

$$= \frac{3}{2} \times \frac{7}{1} = \frac{21}{2} = 10\tfrac{1}{2}$$

or $$15\% = \frac{15}{100} = 0.15$$

Therefore $\quad$ 15% of $70 = 0.15 \times 70 = 10.5$

You must be careful when dealing with percentages since the percentage refers to a fraction of a given quantity, and if the size of this quantity changes so does the percentage value. This is best illustrated by using an example.

Suppose you treat a tray of 200 plants with a weedkiller and 60% die. You treat the remaining plants with a second dose of weed killer and 25% of those remaining die. What percentage has been killed? It is tempting to say 60% + 25% = 85% so 85% have been killed, but this is incorrect.

The first treatment kills 60% of the plants:

$$60\% \quad of \quad 200 = \frac{60}{100} \times \frac{200}{1} = 120$$

Since $200 - 120 = 80$, this means 80 plants remain alive.

After the first dose of weedkiller 80 plants remain and 25% of these are killed by the second dose:

$$25\% \text{ of } 80 = \frac{25}{100} \times \frac{80}{1} = 20$$

so in total $(120 + 20) = 140$ plants have been killed out of the original 200 and as a fraction this is represented by $\frac{140}{200}$ or the equivalent fraction $\frac{7}{10}$. This can be converted to a percentage using the method shown in Example 2.18:

$$\frac{7}{10} \times \frac{100}{1} \% = 70\%$$

As can be seen, when a value is changing in increments (i.e. stages) you cannot simply add or subtract the percentage changes to get the overall percentage change. In the above example the first dose killed 60% and the second 25% of the remainder but in total 70% of the plants were killed, not 85%

The effect of incremental changes in terms of percentages is further illustrated in Example 2.21.

If a value is changing in increments, you cannot simply add or subtract the percentage changes to get the overall percentage change

Example 2.21

A tree measures 5.3 m and over a year its height increases by 10%.

(a) *What is the new height?*

(b) *At the end of the year the tree is topped to decrease its height by 10%. What is the height now?*

(a) $\dfrac{10}{100} \times 5.3 = 0.53\,m$ *so 10% growth will give a height of*

   $5.3 + 0.53 = 5.83\,m$

(b) $\dfrac{10}{100} \times 5.83 = 0.583$ *so if the new height decreases by 10% we have*

   $5.83 - 0.58 = 5.25\,m$

Notice that a 10% increase followed by a 10% decrease does not return you to the starting point. The tree height can be thought of as a variable, h, that is increasing as the tree grows. In Example 2.21 the calculation gives $\frac{1}{10}$th of h in part (a) and in part (b), but because the second calculation used a larger value for h we get a bigger number when looking at $\frac{1}{10}$th of the total. This illustrates the point that if a value is increased by a set percentage, and then this new value is decreased by the same percentage, you do not return to your starting value since you are looking at fractions of a varying total.

Worked examples 2.4

(a) $\frac{7}{10}$ of a sample was used. What percentage remains?

(b) Express the following as percentages of a total:
  (i) $\frac{3}{4}$ (ii) $\frac{2}{3}$ (iii) $\frac{6}{12}$ (iv) $\frac{9}{17}$ (v) $\frac{13}{14}$

(c) Evaluate the following:
  (i) 20% of 80 (ii) 35% of 22 (iii) 83% of 16
  (iv) 12% of 93.

2.8 Ratios

Ratios provide a means of expressing proportions or fractions. For example, you may make up a solution of three parts methanol to one part chloroform. This mixture is often used for extracting lipids from biological membranes. In total you have four parts, three of which are methanol and one of which is chloroform. Both chloroform and methanol are liquids so the final volume is $\frac{3}{4}$ methanol and $\frac{1}{4}$ chloroform. This ratio can be written as:

methanol : chloroform in the ratio 3 : 1

When written in this way the sum of the values gives the total number of parts, with each individual number representing the fraction of the total that is assigned to the corresponding component. If you wanted 100 ml of methanol : chloroform in the ratio 3 : 1 you would therefore add $(\frac{3}{4} \times 100 = 75)$ ml methanol to $(\frac{1}{4} \times 100 = 25)$ ml chloroform.

To calculate the ratio, take the smallest number and divide all the amounts by this value (Example 2.22).

Example 2.22

Given: 10 g *of* A; 5 g *of* B; 15 g *of* C
so the ratio of A:B:C *is* $\frac{10}{5} : \frac{5}{5} : \frac{15}{5}$ *or* A : B : C *in the ratio* 2 : 1 : 3

It is usual to try to give ratios in integer values, although fractions can be used. It may be that your smallest quantity will not divide into the other values. In this case you can use prime factorisation to try to find **the highest common factor** for the numbers concerned or, if the quantities cannot be simplified, express the ratio with the original values.

The highest common factor is the biggest number that will divide exactly into all the numbers of interest, and can be obtained by multiplying together the prime factors which are common to the numbers concerned.

Example 2.23

$$84 = 2 \times 42 \qquad and \quad 210 = 2 \times 105$$
$$= 2 \times 2 \times 21 \qquad\qquad\quad = 2 \times 3 \times 35$$
$$= 2 \times 2 \times 3 \times 7 \qquad\qquad = 2 \times 3 \times 5 \times 7$$

Prime factors common to both 28 *and* 210 = {2, 3, 7}
The highest common factor = 2 × 3 × 7 = 42.

Example 2.24

Given: 8 g *of* A; 24 g *of* B; 6 g *of* C.
These all have the highest common factor 2
The ratio A : B : C *is* $\frac{8}{2} : \frac{24}{2} : \frac{6}{2}$ *or* 4 : 12 : 3.

Example 2.25

Given: 11 g *of* A; 2 g *of* B; 13 g *of* C
This ratio cannot be simplified:
Therefore the ratio A : B : C *is* 11 : 2 : 13.

Ratios are often used in biology to describe dilutions, and some students are unclear about how to deal with these. For example, you may be asked to prepare a 1 in 2 (written as 1 : 2) dilution. In this instance the instruction is saying 'Take one part of solution and add two parts of whatever you are diluting it with.' For example, if you have 1 ml of protein in phosphate buffer, to make a 1 : 2 dilution you would take 1 ml of protein solution and add 2 ml of phosphate buffer. Notice, therefore, that your final volume is now 3 ml, i.e. the volume has increased three-fold so the solution has been diluted three-fold. This highlights the fact that if the dilution is expressed in parts or as a ratio, this tells you what size of fractions to combine, but if it is expressed in terms of a dilution factor this tells you how many-fold the final volume must be increased.

Worked examples 2.5

(a) A, B, C and D are all liquids. You require a mixture with a final volume of 100 ml using the following compounds in the ratios given. What volumes of each are required?

(i)   A : B : C in a ratio of 1 : 2 : 2

(ii) A : B in a ratio of 1 : 1

(iii) A : B : C : D in a ratio of 1 : 4 : 3 : 2

(iv) B : C : D in a ratio of 2 : 1 : 3

(b) I have the following amounts of A, B and C. Express these amounts in the simplest ratio possible.

(i)   30 g of A, 5 g of B, 25 g of C

(ii) 0.5 g of A, 1.5 g of B

(iii) 13 g of A, 6 g of B, 3 g of C

(iv) 15 g of A, 6 g of B, 12 g of C

(c) There is 2 ml of stock solution. How much water would be added to give:

(i)   a 1 : 2 dilution;

(ii) a two-fold dilution?

Summary

Rational numbers are a sub-set of real numbers denoted by the symbol $\mathbb{Q}$ and represented in the form:

$$\frac{p}{q} \quad (q \neq 0)$$

where p and q are integers. If a number cannot be represented in the above form it is said to be an irrational number. Fractions are represented by p/q and if p is less than q this is termed a proper fraction; but if the reverse is true it is an improper fraction and can be represented in a mixed form. If the denominator and numerator have a common factor, this can be found by using prime factorisation and the common factors can cancel to give an equivalent fraction. Equivalent fractions are always formed if both the numerator and denominator are multiplied by the same constant, but this is not true if a constant is added to both the numerator and the denominator, or if the fraction as a whole is multiplied by a constant.

To add or subtract fractions, the denominators must be made equal. This can be achieved by finding the lowest common multiple of the denominators involved. For multiplication the denominators are multiplied together, as are the numerators, to give the resultant fraction. Division proceeds in the same way as multiplication but the dividing fraction must first be inverted.

Percentages are simply fractions of 100; but it must be remembered that if a value changes in increments and these changes are measured as percentages, then you cannot simply add the percentage changes together to find the overall percentage change. Fractions of a whole can be represented in the form of a ratio which, where possible, should be represented in its simplest form by division by the highest common factor.

End of unit questions

1. Evaluate the following:

    (a) $\frac{1}{2} + \frac{5}{7}$ (b) $\frac{2}{6} - \frac{1}{4}$ (c) $\frac{6}{7} \times \frac{2}{3}$ (d) $\frac{1}{9} \div \frac{4}{3}$

2. What percentage of the whole do A, B and C form if combined in the following ratios?

    (a) A : B : C in the ratio 2 : 5 : 1

    (b) A : B : C in the ratio 3 : 7 : 14

3. A sapling is 1.3 m tall. In one week it grows by 8% and the second week its height increases by a further 3%.

    (a) What is the height after two weeks?

    (b) What is the percentage increase after two weeks?

4. A farmer uses 70% of his land for agricultural purposes. With this 70% he grows corn : wheat : barley in a ratio of 3 : 1 : 5.

    (a) What percentage of his land is dedicated to each of these crops?

    (b) If his farm is 200 acres, what area is used for each?

5. A patient is given a chemotherapeutic drug. Over a one-day period 40% is secreted and 28% of that remaining is metabolised. What percentage actually remains? Express these values as a ratio of secreted : metabolised : remaining.

6. In thin-layer chromatography a mobile phase moves up a thin layer of silica on a glass plate. The components in the sample are drawn up the plate by the mobile phase and the distance moved depends on each component's relative affinity for the mobile and solid phases. You wish to make 250 ml of mobile phase containing chloroform, methanol and water in the ratio of 65 : 35 : 4. How much of each must be combined to produce the 250 ml?

7. DNA is composed of four nucleotides, each of which contains a phosphate group. The nucleotides can therefore be purchased containing radioactive phosphate to allow you to produce radioactive oligonucleotides which can be detected on film. The activity of the radionucleotide is measured in becquerels (Bq) and can be determined in a scintillation counter. It is known that every 14.3 days half the sample will decay, thus becoming non-radioactive. Your sample initially contains 11 226 Bq of material. After 14.3 days this has halved

to 5613 Bq. After a further 14.3 days the activity has halved again to give 2806.5 Bq.

(a) How long does it take the sample to decay to 6.25% of its original activity?

(b) What percentage of the sample has decayed after 114.4 days? How much is left in Bq?

8. Absolute error was defined in Section 1.3. This is often represented as a percentage of the total, in which case it is termed relative error. The relative error is obtained by dividing the modulus of the error by the true value being measured and converting the fraction to a percentage. A bacterial cell is known to measure 3 µm. In a practical exam a group of students try to measure the length of the bacterium using a graticule. The students' answers had relative errors of (i) 3%, (ii) 10%, (iii) 8%, (iv) 15% and (v) 1%. What measurements did they record?

9. Proteins are composed of amino acids. The peptide hormone, insulin (bovine) contains a range of amino acids, some of which are shown below as a percentage of the total amino acid content. The measurements were made from 0.5 g of sample. Complete the table.

Amino acid	% (w/w)	Amount in 0.5 g (g)
Alanine	4.6	0.023
Arginine	3.1	
Glycine		0.026
Leucine		0.068
Valine		0.097

10. Fatty acid composition can be analysed using gas chromatography. During the preparation of the sample, material can be lost, so an internal standard is added of known concentration. This standard is usually a fatty acid which is not found within the sample. The amount of fatty acid is recorded as a peak and the area under the peak is proportional to the amount of fatty acid present. Since you know the concentration of your standard you can compare the unknown peaks with the standard in the form of a ratio and calculate the concentration of the unknown values. For example, a standard is 20 µM and the corresponding peak has area 2 cm^2. A palmitic acid peak in the same sample has area 1 cm^2. The concentration of palmitic acid is therefore half that of the standard, i.e. 10 µM.

For a 15 µM internal standard the following data were obtained:

ratio of standard : myristic acid : palmitic acid : oleic acid = 7 : 3 : 8 : 12

What are the concentrations of the three fatty acids in the sample?

3 Basic Algebra and Measurement

3.1 Introduction

Scientists spend much time interpreting data and trying to find the relationship between various factors. When a relationship is discovered it may be expressed in a 'general form' which can be used by other workers. This general form of the relationship may represent quantities by symbols or letters. For example, t is often used to represent time. The manipulation of symbols is termed **algebra**, and **algebraic expressions** are simply equations containing letters or a mixture of letters and numbers.

It is important that any algebraic terms are defined not only with respect to the quantity they represent but also with respect to the standard against which they are being measured, e.g. 't represents time (seconds)'. In this section we will consider the importance of units. It is essential that when numerical values are used they are assigned the correct unit. Many students perform calculations but then neglect to express the answer in the correct form. Without the correct units answers are useless, since other investigators do not know what has been measured. You should therefore understand the meaning of the units being used and be able to express your answers correctly in terms of these units.

The objectives of this chapter are:

(a) to introduce the importance and concept of units;
(b) to introduce algebraic notation;
(c) to provide examples of algebraic manipulation;
(d) to provide experience of transposing (i.e. rearranging) formulae.
(e) to introduce inequalities

3.2 Measurement

The magnitude or size of any quantity can only be measured in relation to a given standard. For example, temperature can be measured using the Celsius scale. On this scale

0 °C is defined as the temperature of ice in equilibrium with water under standard pressure. 100 °C is defined as the temperature of water in equilibrium with steam under standard pressure. When you measure the temperature in degrees Celsius you are recording the temperature relative to these points. It can be seen that you must therefore report not only the value recorded for the temperature but the units of measurement, since the units tell other workers against what standard reference point the quantity is being measured; without them the quantity is meaningless. The quantities most often used in life sciences measure dimensions (length, area, volume), mass, time and temperature. Each of these factors has a range of units associated with it: for example, temperature has been described in terms of the Celsius scale but can also be measured in Kelvin units or degrees Fahrenheit. All three scales are completely different since they measure the quantity (temperature) relative to different standards.

Example 3.1

$$Freezing\ point\ of\ water = 0\ °C$$
$$= 32\ °F$$
$$= 273.15\ K$$

Within science the Système International d' Unités or SI system has been adopted. This is an internationally agreed form of measure which assigns basic or primary units to the seven physical quantities listed in Box 3.1.

Box 3.1 **SI base units.**

Quantity	SI unit	Symbol
Length	metre	m
Mass	kilogram	kg
Time	second	s
Electric current	ampere	A
Thermodynamic temperature	kelvin	K
Luminous intensity	candela	cd
Amount of substance	mole	mol

These invariant primary units are used to define a variety of derived units. Commonly occurring derived units within the life sciences are listed in Box 3.2.

Box 3.2 **SI derived units.**

Quantity	SI unit	Symbol	Definition
Energy	joule	J	$m^2\,kg\,s^{-2}$
Force	newton	N	$m\,kg\,s^{-2}$
Pressure	pascal	Pa	$m^{-1}\,kg\,s^{-2}$
Power	watt	W	$m^2\,kg\,s^{-3}$
Electric charge	coulomb	C	$A\,s$
Electric potential difference	volt	V	$m^2\,kg\,s^{-3}\,A^{-1}$
Electric resistance	ohm	Ω	$m^2\,kg\,s^{-3}\,A^{-2}$
Illumination	lux	lx	$m^{-2}\,cd\,sr$
Frequency	hertz	Hz	s^{-1}

It is worth noting from Box 3.2 that when units are named after people they are written in full with lower-case letters, but when represented by a symbol this tends to be a capital letter (Example 3.2).

Example 3.2

$$1 \text{ newton} = 1\,N$$
$$1 \text{ pascal} = 1\,Pa$$

Whenever you are using units the number should be separated from the unit by a space, the unit should be singular and there is no full stop after the unit (Example 3.3).

Example 3.3

3 metres *is written as* 3 m, *not* 3m *or* 3 ms

When two or more units are combined to form a derived unit a space is left between each unit, but there is never a space between a prefix (Chapter 4) and the symbol to which it applies. Example 3.4 demonstrates this.

Example 3.4

metres per second *is given by* $m\,s^{-1}$
1 millisecond (*the prefix milli indicates one-thousandth*) = 1 ms

Notice in Example 3.4 that 'per' means divide and is represented by a negative superscript. For example, acceleration in metres per second squared is $m\,s^{-2}$. This is covered in more detail in Chapter 4 but the convention should be noted and whenever possible workers should adhere to this notation rather than using a slash (Example 3.5).

Example 3.5

$$\text{metres per second} = \text{m s}^{-1} \text{rather than m/s}$$

This convention is preferred because many texts use a solidus (slash) to separate a symbol from its units, for example if time in seconds is represented by the letter t, then graphs and tables could contain the heading t/s to indicate that the units are seconds.

The following key rules should be followed when using units:

(a)   All quantities should be represented by a number and a unit. The choice of units must be consistent so that in any piece of work you use the same units throughout for any given quantity. Whilst SI units should be used wherever possible, sometimes this is not feasible; for example if you are measuring CO_2 evolution from a plant over a 24-hour period it would be better to use hours rather than seconds.

(b)   Only quantities which have the same units can be added or subtracted so for example, you can not subtract a mass (kg) from time (s).

(c)   There are two instances where units are not used: the first is in the case of ratios (Section 2.8), but only when the ratio is composed of two quantities with the same units. In this case the units cancel (Example 3.6).

Example 3.6

$$\frac{3\,\text{m}}{6\,\text{m}} = 0.5 \quad but \quad \frac{3\,\text{m kg s}^{-2}}{6\,\text{kg}} = 0.5\,\text{m s}^{-2}$$

The second case is that of logarithms, which is covered in Chapter 7.

3.3 Algebraic notation

As referred to in the Introduction, algebraic notation refers to the practice of using a letter to represent a quantity. Although certain quantities are associated with given symbols, it is up to the users to choose whatever letter or symbol they want to represent a quantity. The important point is that the symbol is fully defined and that where appropriate the definition includes units.

Example 3.7

$$t = \text{time (seconds) } or \; t/\text{s}$$
$$l = \text{length (metres) } or \; l/\text{m}$$

It is important that once a symbol is defined it is used consistently to represent the same quantity throughout that piece of work. It should be noted that changes in case can also be used to differentiate between quantities so for example, t would not be considered the same as T.

A symbol can be used to represent a quantity that varies such as the example of time given above, and it is then said to represent a **variable**. If the symbol represents a fixed value, then this is termed a **constant**. As well as using the character set associated with the English alphabet it is common practice to use Greek letters. For example, the symbol for pi (π) is usually used to represent a constant which is approximated by $\frac{22}{7}$. The rules of addition, subtraction, multiplication and division that were discussed in Chapter 1 also apply to algebraic expressions. This means that the priority of operations (Box 1.2) remains the same and the commutative and associative laws (Boxes 1.3 and 1.4) can be applied.

3.3.1 Addition

This is usually referred to as a sum, so Example 3.8 refers to the sum of a and b, where a and b represent two undefined quantities.

Example 3.8

$$a + b = b + a \quad (Commutative \; law \; of \; addition)$$

3.3.2 Subtraction

This may be referred to as a difference, so Example 3.9 represents the difference of a and b.

Example 3.9

$$a - b = a + (-b) = (-b) + a$$

Since a and b represent two different quantities, they are represented by different symbols, and with both addition and subtraction the expressions cannot be simplified any further because you cannot add or subtract different quantities. If the expression contained the same quantities, then the sum or the difference can actually be evaluated.

Example 3.10

$$3a - a = 2a$$

Example 3.11

$$5b + a - b = 5b - b + a \quad (Law\ of\ association)$$
$$= 4b + a$$

3.3.3 Multiplication

This is termed a product and can be written in several different ways $(a \times b = ab = a.b)$. The product of a and b is shown in Example 3.12.

Example 3.12

$$a \times b = b \times a \quad (Commutative\ law\ of\ multiplication)$$

3.3.4 Division

The use of division provides an algebraic fraction which can be treated in the same way as the fractions covered in Chapter 2. The top line is termed the **numerator** and the bottom line is the **denominator**. The term **quotient** is used to describe division, so in Example 3.13 a/b is the quotient of a and b.

Example 3.13

$$a \div b = a/b = a \times \frac{1}{b}$$

3.3.5 Brackets

You may find that the algebraic expression contains brackets; to simplify the expression it may be necessary to removethem. Whatever quantity or symbol is found adjacent to the left-hand side of the brackets must multiply the contents of the brackets and this includes the addition or subtraction sign. It is necessary at this stage to apply the rules for negative numbers in Box 1.1. Observe the two expressions in Examples 3.14 and 3.15. These expressions are completely different, yet with Example 3.14 many students fail to multiply by the negative sign when they remove the brackets, thus incorrectly giving $a - b + c$.

Example 3.14

$$a - (b + c) = a - b - c$$

Example 3.15

$$(a - b) + c = a - b + c$$

Worked examples 3.1

Simplify the following where possible:
(i) $t - (2t + c)$ (ii) $p + c - p$ (iii) $xy + 2x - y + 4xy$
(iv) $z + (t - c)$ (v) $-2(3 - y)$.

3.4 Substitution

Substitution is the process by which symbols within an algebraic expression are replaced by numerical values. If you have performed any algebraic manipulation it is often useful to substitute the symbols for simple numbers to ensure that the manipulated expression still gives the same answer as the original expression. This could be done for Example 3.14 as shown in Example 3.16.

Example 3.16

$$a - (b + c) = a - b - c$$

Let $a = 5, b = 3$ and $c = 1$

After manipulating an algebraic expression, use substitution to check the answer

$$a - (b + c) \quad \text{(using substitution)}$$
$$= 5 - (3 + 1)$$
$$= 5 - 4 \quad \text{(brackets first)}$$
$$= 1$$

Also $a - b - c = 5 - 3 - 1 = 1$

By using substitution it would appear that the removal of the brackets has not affected the expression, so the manipulation is correct.

3.5 Factorising simple formulae

As discussed in Chapter 2, factorising involves expressing a number in terms of a product. In Example 2.4, nine is expressed in terms of its factor three $\{9 = 3 \times 3\}$. If an algebraic expression has more then one term but the terms contain a common factor, then the common factor can be removed (Example 3.17).

Example 3.17

$$3a + 9b \qquad \text{*Both terms (i.e. 3a and 9b) contain a common factor (i.e. 3)*}$$
$$= 3(a + 3b) \qquad \text{*Divide by the common factor and take it outside the brackets*}$$

Notice that anything placed alongside the bracket in this way must multiply everything in the brackets (Example 3.18).

Example 3.18

$$3(3x + y) = 3 \times (3x) + 3 \times (y)$$

Common factors could include symbols as well as numbers, as can be seen in Example 3.19.

Example 3.19

$$xy - y = y(x - 1)$$

It is useful to be able to find the largest number which will divide all the factors you are interested in. You can use prime factorisation to help find the **highest common factor** of two or more numbers (Section 2.8). If you know the highest common factor, then sometimes this can be used to simplify equations using the **distributive law**. This states that:

instead of multiplying two numbers by a common factor, you can add the numbers and then multiply the sum by the common factor (Box 3.3)

Box 3.3

$$ab + ac = a(b + c)$$

The distributive law can be used to simplify a range of operations, especially where the same calculation is repeated a number of times. For example, consider converting degrees Fahrenheit to degrees Celsius. To convert to Celsius the following calculation must be performed, where F represents the reading in Fahrenheit:

$$((F \times 5) - 160) \div 9$$

If you have many readings to convert this is a laborious task and due to the number of operations it can be prone to error. This is simplified in Example 3.20

Example 3.20

Notice that:

$$(F \times 5) - 160$$
$$= 5F - 160$$
$$= 5(F - 32) \qquad using\ the\ distributive\ law$$

so $((F \times 5) - 160) \div 9 = 5(F - 32) \div 9$
or $\frac{5}{9}(F - 32)$ *which is approximately* $0.56(F - 32)$

Worked examples 3.2

Where possible find the highest common factor of:
(i) 18 and 96   (ii) 9, 35 and 27   (iii) 44, 220 and 66
(iv) 90 and 126   (v) 54 and 135.

3.6 Algebraic fractions

An algebraic fraction is a fraction in which either the numerator or denominator (or both) contains an algebraic expression. These fractions can be simplified by cancelling common factors in the same way as numerical fractions can be simplified (Example 3.21).

Example 3.21

$$\frac{x}{xy - 3x}$$

$$= \frac{x}{x(y - 3)} \qquad \textit{Factorise } xy - 3x$$

$$= \frac{1}{y - 3} \qquad \textit{Cancel common factors}$$

3.6.1 Multiplication and division of algebraic fractions

Multiplication and division follow the same rules as numerical fractions. With multiplication, simply multiply the numerators and multiply the denominators (Example 3.22).

Example 3.22

$$\frac{a}{b} \times \frac{c}{d} = \frac{ac}{bd}$$

In the case of division, the dividing fraction should be inverted and then the numerators are multiplied and the denominators are multiplied (Example 3.23).

Example 3.23

$$\frac{a}{b} \div \frac{c}{d} = \frac{a}{b} \times \frac{d}{c} = \frac{ad}{cb}$$

The use of algebraic fractions often occurs when dealing with proportions and is very common when calculating dilutions and concentrations.

3.6.2 Addition and subtraction of algebraic fractions

Addition and subtraction require all the fractions concerned to have the same denominator and the operation proceeds as described in Chapter 2 for numerical fractions. The easiest way to give all the fractions a common denominator is to multiply the denominators together, remembering that if you multiply the bottom of the fraction by a given factor then you must also multiply the top by the same amount to obtain an equivalent fraction (Box 2.2). The process is illustrated in Example 3.24.

Example 3.24

Evaluate the following:

$$\frac{2}{x} - \frac{6}{y}$$

so $$\frac{2}{x} = \frac{2y}{xy} \quad and \quad \frac{6}{y} = \frac{6x}{xy}$$

Therefore $$\frac{2}{x} - \frac{6}{y} = \frac{2y}{xy} - \frac{6x}{xy}$$

$$= \frac{(2y - 6x)}{xy}$$

$$= \frac{2(y - 3x)}{xy}$$

Worked examples 3.3

Simplify the following:

(i) $\dfrac{2ab}{(ab + 3ab)}$ (ii) $\dfrac{3x}{(6 - 18x)}$ (iii) $\dfrac{ab}{(ab + a)}$ (iv) $\dfrac{3a}{6b} \times \dfrac{3b}{a}$

(v) $\dfrac{3}{2} \times \dfrac{t}{7}$ (vi) $\dfrac{2}{(a + b)} - \dfrac{6}{b}$ (vii) $\dfrac{3xy}{t} + \dfrac{7xy}{2m}$

3.7 Transposing formulae

Transposing formulae simply involves rearranging the symbols. In Example 3.25 the symbol x is said to be the **subject** of the equation since it appears alone on one side of the equality.

Example 3.25

$$x = 2y + a$$

If you are rearranging an equation there is only one key rule to apply. Whatever you do to one side of the equation, you do to the other. For example if the equation in example 3.25 is transposed to make y the subject, then the following operations need to be performed:

(a)   subtract a from both sides:

$$x = 2y + a \quad so \quad x - a = 2y + a - a$$
$$= 2y$$

(b)   divide both sides by two:

$$x - a = 2y \quad so \quad \frac{x-a}{2} = \frac{2y}{2}$$

$$= y$$

(c)   it is usual to write the subject on the left, so reverse the equation:

$$y = \frac{(x-a)}{2}$$

Notice in the above example that if one quantity is divided by two, then all the quantities present must be divided by two, otherwise one side of the equation would change. This is emphasised in Example 3.26, where the equation is transposed, by two different strategies, to make y the subject.

Example 3.26

(a)
$$x = 2y + a - 2c$$

$$x - a = 2y - 2c + a - a \quad Subtract\ a$$

$$x = 2y - 2c$$

$$\frac{x-a}{2} = \frac{2y-2c}{2} \qquad Divide\ by\ 2$$

$$= y - c$$

$$\frac{(x-a)}{2} + c = y - c + c \qquad Add\ c$$

$$= y$$

$$y = \frac{(x-a)}{2} + c \qquad Put\ subject\ on\ left$$

(b)
$$x = 2y + a - 2c$$

$$x - a = 2y - 2c + a - a \quad Subtract\ a$$

$$= 2y - 2c$$

$$x - a + 2c = 2y - 2c + 2c \qquad Add\ 2c$$

$$= 2y$$

$$\frac{(x-a)}{2} + \frac{2c}{2} = \frac{2y}{2} \qquad Divide\ by\ 2$$

$$\frac{(x-a)}{2} + c = y$$

$$y = \frac{(x-a)}{2} + c \quad Put\ subject\ on\ left$$

At times the value of interest may be enclosed in brackets, in which case the brackets need to be removed. This is illustrated in Example 3.27, where the formula is transposed to make x the subject. Again two methods are shown.

Example 3.27

(a)
$$y = 9(x - 2)$$
$$\frac{y}{9} = \frac{9(x - 2)}{9} \quad \textit{Divide by 9}$$
$$= x - 2$$
$$\frac{y}{9} + 2 = x - 2 + 2 \quad \textit{Add 2}$$
$$= x$$
$$x = \frac{y}{9} + 2 \qquad \textit{Put subject on left}$$

(b)
$$y = 9(x - 2)$$
$$y = 9x - 18 \qquad \textit{Remove brackets}$$
$$y + 18 = 9x + 18 - 18 \quad \textit{Add 18}$$
$$= 9x$$
$$\frac{y}{9} + \frac{18}{9} = \frac{9x}{9} \qquad \textit{Divide by 9}$$
$$x = \frac{y}{9} + 2 \qquad \textit{Put subject on left}$$

With some equations it is not quite so easy to alter the subject, since it might occur more than once or be part of a product. In this instance the first step involves isolating the factor of interest (Example 3.28).

Example 3.28

(a) *Make* y *the subject:*
$$x - xy = 7$$
$$-xy = 7 - x \qquad \textit{Subtract x}$$
$$-y = \frac{7 - x}{x} \qquad \textit{Divide by x}$$
$$= \frac{7}{x} - 1$$
$$y = -\frac{7}{x} + 1 \qquad \textit{Multiply by} -1$$
$$y = 1 - \frac{7}{x}$$

(b) $x - xy = 7$

$$\frac{x}{x} - \frac{xy}{x} = \frac{7}{x} \qquad \textit{Divide by } x$$

$$1 - y = \frac{7}{x} \qquad \textit{Subtract } 1$$

$$-y = \frac{7}{x} - 1$$

$$y = -\frac{7}{x} + 1 \qquad \textit{Multiply by } -1$$

$$y = 1 - \frac{7}{x}$$

In Example 3.28 it is relatively simple to obtain y on its own since we can remove x from the product xy, by dividing by x. If it was decided to make x the subject this would be a little more difficult since we must isolate x. This can be done by factorisation since x is a common factor of both x and xy, as seen in Example 3.29.

Example 3.29

Make x *the subject:*

$x - xy = 7$ *x is a common factor and therefore can be removed*

$x(1 - y) = 7$ *Divide by* $(1 - y)$

$$x = \frac{7}{(1 - y)}$$

Worked examples 3.4

In the following cases transpose the formulae to make x the subject:

(i) $y = \dfrac{2}{x}$ (ii) $y = \dfrac{7}{x - 3}$ (iii) $y = (x - 6) - 2$

(iv) $2 = 3xy$

3.8 Inequalities

So far we have dealt with simple equalities such as

$$2x = 4$$

This can be read as $2x$ is numerically equal to four. There are occasions in life sciences when you may want to express the relationship between factors in the form of an equation but it may be that the left- and right-hand sides of the equation are not so clearly defined. This can lead to an inequality which uses the symbols listed in Box 3.4

Box 3.4

$a > b$	a is strictly greater than b
$a < b$	a is strictly less than b
$a \geq b$	a is greater than or equal to b
$a \leq b$	a is less than or equal to b

Inequalities can be solved in the same way as equalities by using the principle that whatever you do to one side of the expression, you do the same to the other.

Example 3.30

$$2x - 3 < x + 1$$
$$2x < x + 4 \qquad Add\ 3$$
$$x < 4 \qquad Subtract\ x$$

If an inequality is divided or multiplied by a negative number, the sign of the inequality must be reversed

The solution to the inequality in Example 3.30 is therefore that x must be less than four. There is one key rule that must be remembered when dealing with inequalities: if you multiply or divide by a negative number, the direction of the inequality changes. This is illustrated in Example 3.31.

Example 3.31

(a) $-1 < 2$ *Multiplying by* -1 *reverses the inequality*
            $1 > -2$

(b) $2 > -6$ *Dividing by* -2 *reverses the inequality*
        $-1 < 3$

A possible problem can arise when the equality contains power terms since this can lead to more than one solution. This is discussed in Chapter 8.

3.8.1 Intervals

Inequalities can be used to express intervals. There are three types of interval, which are described below.

3.8.1.1 *Closed intervals*

A closed interval is one for which the end points are included in the interval. It is represented by square brackets.

Example 3.32

The interval from two to six inclusive of two and six is given by: [2, 6]

$x \in [2, 6]$ *means x is a member of the interval* [2, 6]

so $2 \leq x \leq 6$

3.8.1.2 Open intervals.

In this case the end points are stated but are not included in the interval. This can be recorded using round brackets or reversed square brackets (Example 3.33).

Example 3.33

The interval from two to six excluding these values is given by (2, 6) *or*]2, 6[*and*

$$x \in (2, 6) \ \textit{means} \ 2 < x < 6$$

In Example 3.33, x is said to be **strictly greater than** two and **strictly less than** six.

3.8.1.3 Semi-open intervals

These can also be called semi-closed intervals, depending on your viewpoint. In this case one of the end points is included in the interval and the other is excluded (Example 3.34).

Example 3.34

The interval from but excluding two, up to and including six, is given by: (2, 6] *and*

$$x \in (2, 6] \quad \textit{means} \quad 2 < x \leq 6$$

Worked examples 3.5

(a) Use inequalities to describe the following intervals:
     (i) [−1, 3)   (ii) (6, 11)   (iii) (0, 8]   (iv) [4, 5]

(b) Solve the following inequalities for x:
     (i) $x - 3 > 2$ (ii) $6 - x > 4$ (iii) $7 + x \geq 6$

3.9 Applications in biological science

Throughout science, examples of algebra and the need for algebraic manipulation can be seen. In this section a simple example of algebraic manipulation is included to highlight its importance, but further examples will be found in the following chapters and among the end of unit questions. Most standard textbooks will also contain examples of relationships which have been described algebraically.

3.9.1 Equilibrium constants—an example of an algebraic fraction

A simple single substrate reaction can be represented as follows:

$$A \xrightarrow{k_1} P$$

where A is the substrate, P is the product and k_1 is a rate constant. This of course becomes more complex if there is more than one substrate or product:

$$A + B \xrightarrow{k_1} P + Q$$

Unfortunately most of the reactions which occur within the cell do not simply continue in this way. Many reactions in nature are reversible and so, rather than proceed to completion, the reaction reaches equilibrium:

$$A + B \underset{k_{-1}}{\overset{k_1}{\rightleftharpoons}} P + Q$$

where k_{-1} is the rate constant for the reverse reaction. At this point the net rate of reaction is zero because the rate of the forward reaction is equal to that of the reverse reaction, i.e. the product is being used at the same rate as it is being formed. The position of the equilibrium is described by the equilibrium constant K_{eq} which is defined as:

$$K_{eq} = \frac{[P][Q]}{[A][B]} = \frac{k_1}{k_{-1}}$$

where square brackets [ ] represent concentrations in moles per litre.

Worked examples 3.6

(i) The forward rate of reaction (v_f) in the above example is given by:

$$v_f = k_1[A][B]$$

Rearrange the following equation to find the rate of the reverse reaction v_r. (Note that at equilibrium $v_f = v_r$.)

$$\frac{[P][Q]}{[A][B]} = \frac{k_1}{k_{-1}}$$

(ii) The units of the rate constants depend on the rate equation. In the equation for v_f above, the rate measures the number of moles of product formed per minute (i.e. $mol\,litre^{-1}\,min^{-1}$). The concentrations of A and B are measured in moles per litre (i.e. $mol\,litre^{-1}$). Transpose the formulae above to make k_1 the subject. Insert the units into the equation and by cancelling common factors find the units of k_1.

Summary

Algebra involves the use of symbols and letters to express relationships in a general form. If the letter refers to a value that can change it is termed a variable; if it is fixed it is termed a constant. It is important that any symbols used in algebra are clearly defined in terms of the quantity they represent and the standard against which they are measured. It is therefore necessary to always state the units. The units should be chosen from the Système International d'Unités base units and derived units listed in Boxes 3.1 and 3.2. Manipulation of algebraic symbols obeys the same rules as numerical manipulation, so the commutative and associative laws can be applied. Algebraic expressions can be simplified by combining like terms or by removing common factors with the application of the distributive law:

$$ab + ac = a(b + c)$$

If an equation needs transposing to alter the subject, then the transposition should be performed in steps in order to isolate the new subject on one side of the equality.

Whatever operation is performed on one side of the equation must also be performed on the other side to ensure that the equality remains. After manipulating any algebraic expression it is useful to substitute integer values into the original equation and the new equation to ensure that both give the same answer and that the manipulation has not altered the expression.

Relationships can also be expressed using inequalities, which can be manipulated in the same way as equalities, but if the expression is multiplied or divided by a negative number, then the sign of the inequality is reversed. A variable x can be defined as belonging to a given interval using inequalities. If this interval includes the end points it is said to be closed and is denoted by square brackets. If it does not include the end points it is said to be an open interval and this is denoted by round brackets. If one end point is included, the interval is semi-closed or semi-open. The potential intervals are listed below

Box 3.5

$x \in [a,b]$	$a \leq x \leq b$
$x \in (a,b)$	$a < x < b$
$x \in [a,b)$	$a \leq x < b$
$x \in (a,b]$	$a < x \leq b$

End of unit questions

1. (a) In Section 3.9.1 equilibrium was discussed. It is known that:

   $$v_r = k_{-1}[P]$$

   for the reaction described by the equation:

   $$A + B \underset{k_{-1}}{\overset{k_1}{\leftrightarrow}} P$$

   v_r measures the number of moles of product converted to substrate per minute (i.e. $mol\,litre^{-1}\,min^{-1}$) The concentration of P is measured in moles per litre (i.e. $mol\,litre^{-1}$). Transpose the formulae above to make k_{-1} the subject. Insert the units into the equation and by cancelling common factors find the units of k_{-1}.

   (b) Can you directly compare the value for k_{-1} in question 1(a) with the value of k_{-1} for the reaction in Section 3.9.1 where $A + B \leftrightarrow P + Q$? Explain your answer. (*Hint*: derive the units in both cases.)

2. A reaction follows Michaelis–Menten kinetics and can be described by the equation given below; $0.000\,1\,mol\,litre^{-1}$ of substrate are converted to product with an initial velocity (v) of $0.000\,006\,mol\,litre^{-1}\,min^{-1}$. The K_m for the reaction is $0.000\,25\,mol\,litre^{-1}$. [S] denotes substrate concentration in $mol\,litre^{-1}$. What is the maximum velocity (V_{max}) this reaction could reach under these conditions?

   $$v = \frac{V_{max}[S]}{K_m + [S]}$$

3. The molecular weight of a macromolecule such as a protein can be determined by looking at its sedimentation properties—the principle being that if you compare molecules of similar density (e.g. two proteins), then the higher the molecular weight the faster the molecule will sediment. The relationship is described by the Svedberg equation:

   $$\text{Molecular weight} = \frac{RTs}{D(1 - v\rho)}$$

   where $R =$ the gas constant
   $(83\,140\,000\,erg\,mol^{-1}\,degree^{-1})$
   $T =$ the temperature (K)
   $D =$ the diffusion coefficient ($cm^2\,s^{-1}$)

$v =$ the specific volume of the macromolecule $(m^3\,kg^{-1})$

$\rho =$ the density of the solvent $(kg\,m^{-3})$

$s =$ the sedimentation coefficient (s)

At 20 °C human serum albumin has a diffusion coefficient of 0.000 000 6 cm^2 s^{-1} and a sedimentation coefficient of 4.6×10^{-13} s. The density of water at this temperature is 0.998 kg m^{-3}. What is the molecular weight of albumin, assuming a specific volume of 0.74 m^3 kg^{-1}? (Note you can assume 0 °C to be equal to 273 K.)

4. Oxygen uptake can be measured using a Warburg manometer flask. Gas exchange causes a change in the pressure in the manometer which is recorded as movement in the position of the manometer fluid. The greater the distance moved, the greater the pressure change. It is intended to measure oxygen uptake by bacteria. The distance moved by the fluid is multiplied by the manometer constant for oxygen to give the oxygen uptake. The constant must first be calculated using the equation:

$$K = \frac{V_g\left(\dfrac{273}{T}\right) + V_f \times \alpha}{P_0}$$

where V_g is the volume of available gas space in mm^3

T is the temperature in Kelvin

V_f is the volume of fluid in the manometer flask in mm^3

α is the absorption coefficient of the gas in the liquid content of the manometer at temperature T

P_0 is the normal atmospheric pressure expressed in mm of manometer fluid ($=10\,000$ mm manometer fluid)

A manometer flask (23 cm^3 volume) contains 3 cm^3 of bacterial suspension at 37 °C. The uptake of oxygen is to be monitored. The absorption coefficient for oxygen at 37 °C is 0.024.

(a) Evaluate the manometer constant for oxygen in this system.

(b) The manometer constant for nitrogen is 1.767 at 37 °C. Transpose the equation to make α the subject and find the absorption coefficient for nitrogen in this system.

5. Rats were fed on a diet depleted in vitamin B and their weights were measured and compared with controls.

The rats were placed in the following groups, with weights:

(i)   up to but not including 30 g;

(ii) from 30 g up to but not including 35 g;

(iii) 35 g up to but not including 40 g; or

(iv) greater than or equal to 40 g.

If the weight is denoted by w (g):

(a) Define the above intervals using inequalities; and

(b) Write down the intervals for groups (ii) to (iii) using bracket notation.

6. The number of visits made by bees to various plants was recorded over a one-hour period. The observations were grouped as follows: no visits; ten or more visits; or visits in the intervals [1,5) and [5,10). Express these using inequalities.

7. An inhibitor that competes with the substrate for the enzyme's active site is called a competitive inhibitor. The velocity of a reaction in the presence of such an inhibitor is described by the following equation:

$$v_i = \frac{[S]\,V_{max}}{K_m\left(1 + \dfrac{[I]}{K_i}\right) + [S]}$$

where v_i is the initial velocity in the presence of the inhibitor, [S] is the substrate concentration and [I] the inhibitor concentration (both in mol litre^{-1}), V_{max} is the maximum velocity that can be reached, and K_m and K_i can be thought of as binding constants for the substrate and inhibitor respectively.

An initial velocity of 0.000 001 4 mol litre^{-1} min^{-1} is observed for a system where the K_m is 0.000 047 mol litre^{-1}. The V_{max} value was 0.000 022 mol litre^{-1} min^{-1}. The amount of substrate present was 0.000 2 mol litre^{-1} and the amount of inhibitor 0.000 4 mol litre^{-1}. Transpose the equation to make K_i the subject, and find K_i.

4 Powers and Scientific Notation

4.1 Introduction

One problem faced in the life sciences is that of the large range of values required. For example, a scientist measuring the size of bacteria may be working at the level of 0.000 001 m yet if the same scientist were to measure the number of bacteria in 1 ml of culture(s) he may well find numbers in the region of 1 000 000 or more. Writing very large or very small numbers in full is tedious and time-consuming, and can lead to errors. Numbers therefore tend to be written using scientific notation, but to understand this form of expression it is necessary to be familiar with powers. The objectives of this chapter are:

(a) to introduce powers and develop confidence in their use;

(b) to develop an appreciation of power rules and their function;

(c) to show how scientific notation can be used to express a range of numerical values.

4.2 Powers

Within the life sciences there are many instances where a number needs to be multiplied by itself several times. For example, you may be growing bacteria on nutrient agar plates. If you grow the bacteria overnight in liquid culture you will have a dense bacterial suspension which contains millions of bacterial cells. Although all these cells should be the same, some may be different due to contamination or mutation, so for experimental work you may want to pick just one bacterial cell and its offspring.

Some of the culture may therefore be spread on an agar plate to allow the growth of single cells, but if the culture is too dense, even a small volume will contain many bacteria and so the plate will be completely covered with growth. If the culture is diluted and then the same volume is plated out, there are fewer bacterial cells on the plate and those present tend to be isolated. If left, these isolated bacteria will grow and multiply to give rise to small colonies which can be selected and which are known to contain bacteria that have all come from the same parent. The problem is how to obtain just the right number of bacterial colonies. If the culture is too dilute you get nothing, yet if it is too concentrated the colonies will merge together due to the large number of bacteria present. Microbiologists tend to take the overnight culture and repeatedly dilute it by a factor of ten. These dilutions are then plated: and some of them will give rise to plates with the correct number of colonies. This technique is termed a serial dilution and is commonly used in science (Table 4.1).

Table 4.1

Dilution	Dilution factor
0	0
1	10
2	$10 \times 10 = 100$
3	$10 \times 10 \times 10 = 1000$
4	$10 \times 10 \times 10 \times 10 = 10\,000$

Each of the dilutions increases by a factor of ten. The dilution factors can be written using power notation (Example 4.1).

Example 4.1

$$10 \times 10 = 100 = 10^2$$

In this example 10 is termed the **base** and this is the factor which is being multiplied. The superscript '2' is termed the **index** or **exponent** and this tells you the number of times the base multiples itself. The full expression is termed a **power**, so in Example 4.1 we have ten to the exponent two. Here we have used base 10, but any base can be used. For example, let the letter a represent any integer, then

$$a \times a \times a = a^3$$

Your calculator will have a key labelled x^y and this can be used to quickly evaluate numbers which involve powers. To evaluate 3^4 type

$$3 \quad x^y \quad 4 \quad =$$

This should give the answer 81. In the serial dilution of bacteria we used a dilution factor of ten, but in some cases this may dilute the sample too quickly. Suppose you prepare a drug which you want to test for toxicity against tumour cells in tissue culture. The toxicity level could be measured in tissue culture by adding increasing amounts of drug until a given percentage of the cells are dead. In the first experiment you dilute the drug repeatedly by a factor of ten. The toxicity level required is found between the 10^3 and 10^4 dilutions. Once you know what range of drug is required, you may want to decrease the dilution factor to make the toxicity assay more sensitive; for example, you could use doubling dilutions.

Example 4.2

A series of solutions are prepared, each of which is $2 \times$ more concentrated than the last. The starting solution contains $3.29\,\mathrm{g\,ml}^{-1}$. What is the concentration of the seventh solution?

i.e.   Solution 0 $3.29\,\mathrm{g\,ml}^{-1}$
Solution 1 $(3.29 \times 2)\,\mathrm{g\,ml}^{-1}$
Solution 2 $(3.29 \times (2 \times 2))\,\mathrm{g\,ml}^{-1} = (3.29 \times 2^2)\,\mathrm{g\,ml}^{-1}$

The concentration of solution 7 is therefore:

$$3.29 \times 2^7 = 3.29 \times 128 = 421.12\,\mathrm{g\,ml}^{-1}$$

If we return to the example with base 10, it can be seen that decreasing the exponent by a factor of one causes the power term to be divided by the base. This is shown in Example 4.3.

Example 4.3

$$10^5 \div 10 = 10^4 \quad \textit{which is the same as}$$
$$100\,000 \div 10 = 10\,000$$

Each time we divide by the base the exponent decreases by one, so eventually we have the situation in Example 4.4.

Example 4.4

$$10^3 \div 10 = 10^2 \quad \textit{or} \quad 1000 \div 10 = 100$$
$$10^2 \div 10 = 10^1 \quad \textit{or} \quad 100 \div 10 = 10$$

So $10^1 = 10$

It is worth noting from Example 4.4 that anything with an exponent of one is simply itself. This process can be taken a stage further since we can continue dividing by the base and decreasing the exponent by one, as shown in Example 4.5.

Example 4.5

$$10^1 \div 10 = 10^0 \quad or \quad 10 \div 10 = 1$$

So $$10^0 = 1$$

Any base with index zero is taken to be one; this is sometimes used to simplify equations. Suppose you continue in this manner—you would end up with negative exponents, as shown in Example 4.6.

Example 4.6

$$10^0 \div 10 = 10^{-1} \quad or \quad 1 \div 10 = 0.1$$

A negative index means 'divide'

It can be seen that with negative exponents the negative sign is taken to mean 'divide'. That is to say, you are no longer multiplying by the base but you are dividing by the power term, as shown below

$$a^{-5} = \frac{1}{a \times a \times a \times a \times a}$$

It should be understood that an index only refers to the base to which it is attached, so for example

$$2a^3 = 2 \times (a \times a \times a)$$

whereas $$(2a)^3 = 2a \times 2a \times 2a = 2 \times a \times 2 \times a \times 2 \times a$$

$$= 8 \times a \times a \times a \text{ (employing the law of association)}$$

$$= 8a^3$$

It is worth mentioning that if you try to use negative bases you need to apply the rules for negative numbers given in Section 1.1 (Example 4.7).

Example 4.7

$$(-2)^3 = (-2) \times (-2) \times (-2)$$
$$= 4 \times (-2)$$
$$= -8$$

So, in general, the rules given in Box 4.1 apply.

Box 4.1 **Negative bases.**

If m is odd $(-a)^m = -(a)^m$

If m is even $(-a)^m = (a)^m$

The use of power notation is summarised in Box 4.2.

Box 4.2 **Power notation.**

$$a^0 = 1$$
$$a^1 = a$$
$$a^m = a_1 \times a_2 \times a_3 \times \cdots a_m$$

$$a^{-m} = \frac{1}{a^m} = \frac{1}{a_1 \times a_2 \times a_3 \times \cdots a_m}$$

Worked examples 4.1

Evaluate:
(i) 5^3 (ii) 2^{-5} (iii) $(-5)^2$ (iv) $(-2)^5$ (v) $(1.147)^9$
(vi) $(-5.73)^5$.

4.3 Multiplication and division using powers

Consider the calculation shown in Example 4.8.

Example 4.8

$$a^2 \times a^3 = (a \times a) \times (a \times a \times a)$$
$$= a \times a \times a \times a \times a$$
$$= a^5$$
$$= a^{(2+3)}$$

It can be seen in this example that if two power terms with the same base are being multiplied, then to obtain the answer you simply need to add the exponents. This leads to the general rule shown in Box 4.3. There is a similar rule for division which is shown in Example 4.9.

Example 4.9

$$a^4 \div a^2 = \frac{a \times a \times a \times a}{a \times a}$$
$$= a \times a$$
$$= a^2$$
$$= a^{(4-2)}$$

In general, if two powers with the same base are being divided, then you simply subtract the exponents. This is summarised in Box 4.3. It is worth remembering that a negative index means 'divide', so Example 4.9 could be re-written using negative indices as in Example 4.10.

Example 4.10

$$a^4 \times a^{-2} = \frac{a \times a \times a \times a}{a \times a}$$
$$= a \times a$$
$$= a^2 \quad \text{(employing power rules for multiplication)}$$
$$= a^{(4-2)}$$

Box 4.3 Power rules: multiplication and division

(i) $a^m \times a^n = a^{(m+n)}$

(ii) $a^m \div a^n = a^{(m-n)}$

These rules give us another way of looking at the effect of negative indices, since:

$$\frac{1}{10^m} = 10^0 \div 10^m$$
$$= 10^{(0-m)} \quad \text{(employing power rules)}$$
$$= 10^{-m}$$

These rules can be used as a powerful tool for simplifying calculations, as can be seen in Example 4.11.

Example 4.11

$$a^2 \times a^5 \times a^{-3} \div a^2 = a^{(2+5+(-3))} \div a^2$$
$$= a^4 \div a^2$$
$$= a^{(4-2)}$$
$$= a^2$$

Worked examples 4.2

(a) Evaluate:
(i) $2^2 \times 2^2$ (ii) $3^3 \times 3^{-3}$ (iii) $2^2 \times 2^4 \div 2^{-3} \times 2^{-4}$
(iv) $3^2 \div 3^5$ (v) $106^{11} \div 106^8$

(b) Simplify:
(i) $6^2 \times 6^9$ (ii) $z^4 z^{-3} z^2$ (iii) $cc^4 c^{-9} c^4$
(iv) $a^2 \times a^{-3} \div a^4$ (v) $c \div c^{-4} \times c^2 \times 1$.

4.4 Powers of powers

You will sometimes find yourself dealing with powers of powers, as shown in Example 4.12. There is also a rule to help you deal with this situation.

Example 4.12

If $a = 2^3$

then
$$a^2 = (2^3)^2$$
$$= 2^3 \times 2^3$$
$$= 2^{(3+3)}$$
$$= 2^6$$
$$= 2^{2 \times 3}$$

When dealing with powers of powers the general rule is simply to multiply them as shown in Example 4.13 and in Box 4.4.

Example 4.13

$$(a^{-4})^2 = a^{(-4 \times 2)} = a^{-8}$$

Box 4.4 **Power rules: powers.**

$$(a^m)^n = a^{mn}$$

Worked examples 4.3

(a) Evaluate: (i) $(2^2)^2$ (ii) $(2^3)^{-3}$ (iii) $(4^{-5})^{-2}$

(b) Simplify: (i) $(a^{12})^3$ (ii) $(e^4)^{-2}$ (iii) $(e^{-3})^{-2}$.

4.5 Fractional indices

Fractional indices represent roots

You may think that fractional powers are uncommon, but they are widely used within science. For example, you will often see the exponent $\frac{1}{2}$ but what does this actually mean? If you try a few values on the calculator you will find that

$$a^{1/2} = \sqrt{a}$$

Indeed $a^{1/n} = \sqrt[n]{a}$ or the nth root of a. The most common fractional exponent is probably that of the square root and this is the example you are most likely to have to deal with.

Fractions obey all the rules described so far; some of these rules are demonstrated in Example 4.14.

Example 4.14

$$(a^{1/2})^2 = a^{1/2} \times a^{1/2}$$
$$= a^{(1/2+1/2)}$$
$$= a^1$$
$$= a$$
$$= a^{(1/2 \times 2)}.$$

4.6 Indices and biology

Why are indices so important in life sciences? The simple answer is that biologists are required to use very large and very small numbers; for example, in a practical class you may need one millionth of a litre, i.e. 0.000 001 litre, or you may be estimating the number of cells in a millilitre of tissue culture media and find the answer is in excess of ten thousand. It can quickly become tedious to write out very large or very small numbers; perhaps you have experienced this when answering the end of unit questions in Chapter 3. With power notation we can abbreviate numbers and make mathematical manipulation simpler by applying the power rules you have just covered. Example 4.15 illustrates how powers are used to simplify large numbers.

Example 4.15

$$10^2 = 10 \times 10 \quad and \ as \quad 200 = 2 \times 100$$
$$so \quad 200 = 2 \times 10^2$$

Instead of writing 200 litre I could therefore write 2×10^2 litre, i.e. (2×100) litres. Here 10^2 is termed the **multiplier**; multipliers can be combined using power rules as shown in Example 4.16.

Example 4.16

$$20\,000 \div 200$$
$$= (2 \times 10^4) \div (2 \times 10^2)$$
$$= (2 \div 2) \times (10^4 \div 10^2)$$
$$= 1 \times 10^2$$

When using power terms to base ten the multiplier tells you where to put the decimal place. Positive powers move the decimal point to the right because the number is getting bigger. Negative powers move it to the left because you are dividing, so the number gets smaller. Example 4.17 demonstrates this.

Example 4.17

10^2 means move the decimal place 2 places to the right:
$2.0 \times 10^2 = 200.0$

10^{-2} means move the decimal place 2 places to the left.
$2.0 \times 10^{-2} = 0.02 = \frac{2}{100}$

Writing numbers with multipliers is very important within science. We often express numbers using **scientific notation**, which means expressing numbers by using multipliers with base ten. It is usual to place the decimal point after the first digit. The multiplier then tells you how many places to move the decimal point to the left or right (see Example 4.18).

In the case of multipliers to the base ten, the index indicates where to move the decimal point

---

Example 4.18

$$3200 = 3.2 \times 10^3$$
$$12\,783 = 1.2783 \times 10^4$$
$$0.000\,45 = 4.5 \times 10^{-4}$$

You will probably also see the term 'order of magnitude' associated with powers to base ten. If any values vary by a factor of ten they are said to be one order of magnitude apart. Variation by a factor of 1000 (i.e. 10^3 or 10^{-3}) would be interpreted as three orders of magnitude apart.

Because we use multipliers so often, some commonly used multipliers have been given names and symbols. Those which are commonly found are listed in Box 4.5.

Box 4.5 **Commonly used multipliers**

Multiplier	Prefix	Symbol	Factor
10^{-1}	deci	d	0.1
10^{-2}	centi	c	0.01
10^{-3}	milli	m	0.001
10^{-6}	micro	μ	0.000 001
10^{-9}	nano	n	0.000 000 001
10^{-12}	pico	p	0.000 000 000 001
10	deca	da	10
10^2	hecto	h	100
10^3	kilo	k	1 000
10^6	mega	M	1 000 000
10^9	giga	G	1 000 000 000
10^{12}	tera	T	1 000 000 000 000

You will have already used a number of these prefixes and symbols, e.g. when dealing with mass (Example 4.19).

Example 4.19

$$1 \text{ kilogram} = 1 \text{ kg}$$
$$= 1 \times 10^3 \text{ grams}$$

Within practical classes you will regularly use milligrams (mg), millilitres (ml) and microlitres (μl).

Worked examples 4.4

(a) Express the following using scientific notation:
     (i) 239   (ii) 0.0036   (iii) 200 × 0.000 003   (iv) 9.73
     (v) 1792 × 0.000 179 2

(b) A solution is prepared and 10 μl are removed for an assay.
     (i) How many litres is this?
     (ii) How many millilitres is it?
     Express your answers in scientific notation.

(c) The 10 μl sample from question (b) had its volume increased 10 000-fold.

     (i)   Express the multiplication factor in scientific notation.

     (ii) What is the new volume?

     (iii) By what order of magnitude has the volume increased?

Summary

A power expression is composed of two parts, the base and the exponent. The exponent is represented by a superscript at the side of the base. A positive exponent represents the number of times the base should be multiplied by itself, so for example 10^4 would mean base ten multiplied by itself four times, i.e. 10 000. A negative exponent means you divide by this number, so 10^{-4} would be $\frac{1}{10\,000}$ or 0.0001. Any base to the power one is itself, and any base to the power zero is one. Exponents can also exist as fractions, where the fraction refers to the root. This is summarised below.

(a) $a^0 = 1$
(b) $a^1 = a$
(c) $a^m = a_1 \times a_2 \times a_3 \times \cdots a_m$

(d) $a^{-m} = \dfrac{1}{a^m} = \dfrac{1}{a_1 \times a_2 \times a_3 \times \cdots a_m}$

The base can be any number, but in calculations involving powers with the same base power rules can be used to simplify the expression. The three rules are summarised below:

(a) $a^m \times a^n = a^{(m+n)}$
(b) $a^m \div a^n = a^{(m-n)}$
(c) $(a^m)^n = a^{mn}$

If the base being used is ten, then when multiplying a number by a power the exponent shows how many places to move the decimal place to the right (positive exponent) or left (negative exponent). When used in this way power expressions are termed multipliers, and these are used to express numbers in scientific notation. Commonly used multipliers can be incorporated into the names of the units by representing the multiplier by one of the symbols listed in Box 4.5.

End of unit questions

1. The circumference of the Earth is about 4.0×10^4 km and that of a bacterial cell is about 3 μm. How many orders of magnitude larger is the Earth compared with the bacterium?

2. The effect of toxicity on mammalian cells can be measured by growing the cells on a plate and measuring their size after a given time interval. This is termed a clonogenic assay. Assume that the cells are spherical then their volume is given by:

$$\text{Volume} = \frac{4}{3}\pi r^3$$

where r is the radius of the cell (m) and π is a constant ($\pi = 3.1416$).

(a) If a typical cell is 4 μm in diameter, what is its volume in metres cubed? Express your answer using scientific notation.

(b) Inclusion of the anticancer drug doxorubicin in the medium decreased the growth by 35%. If growth refers to the cell volume, what is the diameter of a typical cell after treatment?

3. It is estimated that plants produce 9.0×10^{12} kg of molecular oxygen each year. If 32 g of oxygen contains 6.02×10^{23} molecules of the gas, how many molecules of oxygen are produced per year?

4. With any quadruped, the further the front legs are from the rear legs the more the torso can sag. The level of sagging will depend on the effect of gravity and thus the height or thickness of the torso. It is known from physics that there is a limit on the length (l) to height

(h) ratio as given by $l : h^{2/3}$. For example, a young Indian elephant is 153 cm long and of height 135 cm. $h^{2/3}$ is therefore 26.3, giving the ratio 153 : 26.3 or 5.8 : 1. The limit for the ratio appears to be around 7 : 1. Confirm that this holds in the case of a dachshund of length 35 cm and height 12 cm. (Adapted from E. Batschelet (1979) *An Introduction to Mathematics for Life Scientists*, Springer-Verlag, New York.)

5. In question 4 the term $h^{2/3}$ was used. Use power rules to show that:

$$h^{2/3} = \sqrt[3]{h^2}$$

6. In photosynthetic bacteria, chemical energy is produced by the absorption of light. The pigment which absorbs the light energy is p870 and it uses this energy to enter an excited, highly reactive state where an electron can be passed to a neighbouring molecule. In *Rhodopseudomonas sphaeroides* the electron passes from p870 to bacteriochlorophyll in approximately 1 ps and then onto bacteriopheophytin in 4 ps. It then passes to ubiquinone in 2×10^{-10} s. How long does the whole process take?

7. A bacterial cell can transcribe DNA to RNA at approximately 50 nucleotides per second. *Escherichia coli* makes on average 1 in 10^5 mistakes during the transcription process but in a mutant which is unable to correct mistakes the error increases to 1 in 10^2 bases. A protein-coding region is being transcribed and contains 1.062×10^3 bases.

(a) How long does it take to transcribe this region?

(b) In the mutant bacteria how many mistakes can be expected?

(c) Assuming that in the wild-type strain a single mistake is made after 10^5 bases have been transcribed, then how many times would you have to transcribe the gene before expecting one mistake?

8. The ionisation constant for water (K_w) is 1×10^{-14} and is given by:

$$K_w = \text{(hydrogen ion concentration)}$$
$$\times \text{(hydroxyl ion concentration)}$$

If a solution contains 5×10^{-10} mol litre^{-1} of hydrogen ions, what is the concentration of hydroxyl ions?

5 Concentration and Accuracy

5.1 Introduction

Concentration and accuracy are areas which seem to cause difficulty to many students, yet these are probably the most commonly used concepts in the laboratory. To be able to cope with this form of calculation, students need to apply their knowledge of ratios, proportions and percentages from Chapter 2 as well as their understanding of power notation from Chapter 4. Since many students are uncomfortable with concentrations and because these concepts are crucial in many scientific fields, this chapter will cover this area in some detail. The aims of the chapter are:

(a) to ensure students are aware of the differences between volumes, amounts and concentrations;

(b) to ensure that students are aware of the most common methods of measuring concentration and to introduce molarity;

(c) to introduce the concept of significant figures.

5.2 Concentration, volume and amount

A solution is defined in terms of the amount of solute and the amount of solvent

During laboratory work you will have to prepare a number of solutions. If you were told to make 1 ml of salt solution, could you do it? What if the instructions tell you to make a 2 g solution? In both of the examples it is clear that you are not given enough information. To make a solution you need to know:

(a) How much material (**solute**) is needed.

(b) How much liquid (**solvent**) you want the material dissolved in.

This can be summarised using a simple equation showing the relationship between concentration, volume and the amount or quantity of material present (Box 5.1).

Box 5.1

$$\text{Concentration} = \frac{\text{quantity}}{\text{volume}}$$

Although concentration is defined in this way, it can be measured against various standards depending on how the quantity and volume are measured. The most commonly used means of measuring concentration are discussed in this section.

5.1.1 *Percentage weight/volume*

To obtain a 1 g in a 100 ml salt solution, 1 g of salt would be dissolved in a final volume of 100 ml. If solutions are made up in this way concentration can be represented by a percentage, since a percentage gives a fraction of 100. Here the percentage is obtained from a given weight (x in g) in a given volume (100 ml) so this is **termed percentage weight/ volume,** or % (w/v), and is defined in Box 5.2. Two solutions are described in this way in Example 5.1.

Box 5.2

% (w/v) = weight in grams of solute per 100 ml of solution

Example 5.1

$$1\,\text{g}\,(100\,\text{ml})^{-1} = 1\%\,(\text{w/v})\,\text{solution}$$
$$10\,\text{g}\,(100\,\text{ml})^{-1} = 10\%\,(\text{w/v})\,\text{solution}$$

In some texts you may see concentration measured in milligrams per cent (mg%). This is very similar to percentage weight/volume but is defined as the weight of solute in milligrams per 100 ml of solution. This is more commonly found in clinical texts since many measurements such as those of blood sugar and drug concentrations involve small amounts of solute.

5.2.2 *Percentage volume/volume*

Some chemicals, e.g. methanol, are not usually solid and although you could weigh out an amount in grams it would be easier to measure out the amount wanted by volume. In this case you could take 1 ml and add 99 ml of water, so that the methanol forms one-hundredth of the total volume (see Example 5.2). The solution can therefore also be measured as a percentage, but in this case it would be by volume not by weight, so this method gives a **percentage volume/volume** solution. This is defined in Box 5.3.

Box 5.3

% (v/v) = volume in millilitres of solute per 100 ml of solution

Example 5.2

1 ml methanol plus 99 ml water = 1% (v/v) solution

5.2.3 *Percentage weight/weight*

If you mix two compounds of known weight you can have **a percentage weight/weight** solution. For example, the inner membrane of the bacterium *Escherichia coli* contains 75% (w/w) of the phospholipid phosphatidylethanolamine. This means that in every 100 g of membrane lipid isolated, 75 g will be phosphatidylethanolamine. You may see some solutions listed as % (w/w), as in Example 5.3; this is especially popular with acids. A percentage weight/weight solution is defined in Box 5.4.

Box 5.4

% (w/w) = the weight in grams of solute per 100 g of solution

Example 5.3

15 g of salt plus 85 g of water = 15% (w/w) solution

Notice the importance of defining what you mean by 'per cent' when using percentage as a unit of measure. 1 ml of water has a weight of approximately 1 g; therefore a 15% (w/w) solution will contain 15 g of solute in approximately 85 ml of water (total 100 g), but a 15% (w/v) solution would contain 15 g of solute made up to 100 ml. If the solute is denser than water it will occupy less than 15 ml; hence

more than 85 ml of water will have to be added to make up
the 100 ml % (w/v) solution. The % (w/v) solution will therefore contain more water than that of the % (w/w) solution.
Since both contain the same quantity of solute (15 g) the
% (w/v) solution must therefore be more dilute than % (w/
w) solution.

It is very important that volume, quantity of solute and
concentration are kept separate and not confused. Many students seem especially prone to confusing the quantity of
solute and the concentration.

Example 5.4

100 ml *of a* 10% (w/v) *solution of sodium chloride is prepared. Assume that the salt is evenly distributed throughout the solution.* 50 ml *of the solution is removed for an
experiment. What is the amount of salt present, in grams?
What is the concentration of the* 50 ml *sample that is
removed?*

(a) *You have taken half the sample and if the salt is
evenly mixed you have therefore taken half the salt. The
original solution contained* 10 g *so you now have* 5 g.
(b) *The concentration is* $5\,g(50\,ml)^{-1} = 10\,g(100\,ml)^{-1}$
$= 10\%\,(w/v)$, *so the concentration has not changed.*

To calculate the amount of material in a volume removed
from a stock solution it is necessary to use proportions.
This is covered in more detail in Chapter 7, but Example
5.5 demonstrates the method.

Example 5.5

In 100 ml *of water there are* 3 g *of salt. How much is in* 20 ml?
There is 3 g *in* 100 ml *so we have a concentration of* (3/100) g
ml^{-1}.
In 20 ml *there is x g and so the concentration is* (x/20) g ml^{-1}
*If the solution is homogeneous (i.e. evenly mixed) then the
concentration in the* 100 ml *is exactly the same as the con-
centration in the* 20 ml, *i.e.*

$$\frac{3}{100} = \frac{x}{20}$$

$$x = \frac{3}{100} \times 20 \quad \textit{Transpose to make x the subject}$$

$$= 0.6\,g \qquad \textit{(Remember the units)}$$

When working out proportions in this way it is important that the fractions on both sides of the equation have the same units. In Example 5.5 we have:

$$\frac{mass}{volume} = \frac{mass}{volume}$$

Worked examples 5.1

(a) What are the following concentrations in % (w/v)?

(i) 5 g of glucose in a final volume of 50 ml

(ii) 5 g of glucose in a final volume of 75 ml

(iii) 7.5 g of glucose in a final volume of 50 ml

(iv) 7.5 g of glucose in a final volume of 75 ml

(b) Assume that in question (a) you have added 48 ml of water (i.e. the glucose has a volume of 2 ml). 1 ml of water weighs approximately 1 g, so what is the % (w/w) concentration of the solution?

5.2.4 *Moles and molarity*

Maltose is formed by linking two molecules of glucose. Therefore a molecule of maltose has approximately twice the weight of a glucose molecule. If you are given a 30% (w/v) solution of each, would they both contain the same number of molecules? The answer is obviously no, since maltose is twice the weight of glucose: 30 g of maltose will contain 50% fewer molecules than 30 g of glucose.

This should have identified the problem of defining concentration solely in terms of weight. It can be seen that if you make a solution by weight alone, then the concentration of atoms or molecules in a given volume is different for different compounds! Since it is these molecules which are required for reaction, a measure of concentration really wants to show not just how much material is present but how many molecules are present. To overcome this problem we work with **molar concentrations**, where a mole (mol) is defined as:

$$1\,mol = 6.02 \times 10^{23}\,molecules$$

Molarity gives a measure of the concentration of molecules present

Hence we say 1 mol of carbon contains 6.02×10^{23} carbon atoms. This is a quantity and so can be represented in grams. The weight of 1 mole is equivalent to the molecular weight of the molecule in grams (Example 5.6).

Example 5.6

Carbon-12 (^{12}C) *has a molecular weight of* 12

$$1 \text{ mol} = 12 \text{ g} \quad \text{and} \quad 12 \text{ g} = 6.02 \times 10^{23} \text{ carbon atoms}$$

The number of moles can therefore be calculated from the equation shown in Box 5.5.

Box 5.5

$$\text{Moles} = \frac{\text{weight}}{\text{molecular weight}}$$

It should be clear that you do not have a 1 mol solution in the same way that you could not have a 1 g solution—you need a volume. Thus we define molarity as shown in Box 5.6, and a 1 molar (1 M) solution contains 1 mol of solute per litre:

$$1 \text{ M} = 1 \text{ \textbf{molar}}$$
$$= 1 \text{ mol litre}^{-1}$$

Box 5.6

$$\text{molarity} = \frac{\text{number of moles of solute}}{\text{number of litres of solution}}$$

The quantity, moles, and the concentration, molarity, are linked by the equation shown in Box 5.7.

Box 5.7

$$\text{moles} = \text{molarity} \times \frac{\text{volume (ml)}}{1000}$$
$$= \text{molarity} \times \text{volume in litres}$$

Whilst this equation can be used, it is recommended that students are also able perform calculations on moles and molarity from first principles. As a check to make sure that the answer is reasonable, remember that if the amount of material increases or the volume decreases then the concentration has increased. If both the quantity of material and the volume alter in the same proportion then the concentration stays the same (Example 5.7).

Example 5.7

$1\,\mathrm{mg\,litre}^{-1}$ *Decrease both amount and volume by 1000-fold*

$= 1\,\mathrm{\mu g\,ml}^{-1}$

Worked examples 5.2

(i) How much sodium chloride is needed to make 1 litre of a 20 mM solution?

Mol.wt of sodium chloride = 58.5.

(ii) The experiment requires 2.3 ml of 20 mM sodium chloride so I will make 5 ml, not 1 litre. How much sodium chloride is now required?

(iii) What weight of sodium chloride contains 1 μmol?

If a concentration is given in % (w/v) and the molecular weight of the material is known, then it is relatively easy to convert to molarity, as shown in Example 5.8.

Example 5.8

A 3% (w/v) citric acid solution is prepared. The molecular weight of citric acid is 192.

$$3\%\,(\mathrm{w/v}) = 3\,\mathrm{g}\,(100\,\mathrm{ml})^{-1}$$
$$= 30\,\mathrm{g}\,(1000\,\mathrm{ml})^{-1}$$

The number of moles present in 1 litre $= \dfrac{30}{192} = 0.16\,\mathrm{mol}$

Hence the solution has a concentration of approximately 0.16 M.

To perform a similar calculation but starting with a solution whose concentration is in % (w/w), the density of the material needs to be known. The conversion between % (w/w) and molarity is often required when measuring out acids. For example, suppose you want 500 ml of a 0.1 M solution of hydrochloric acid but the acid has been bought as a 27% (w/w) solution.

Rearranging the equation from Box 5.6, we require :

$$0.1 \times \frac{500}{1000} = 0.05\,\mathrm{mol}$$

The molecular weight of hydrochloric acid is 36.5, so we require :

$$0.05 \times 36.5 = 1.83\,\mathrm{g}$$

The stock acid is 27% (w/w), i.e. 100 g would contain 27 g of acid. To obtain 1.83 g of acid we would therefore need:

$$\frac{1.83}{0.27} = 6.78\,\text{g of stock.}$$

6.78 g could be weighed out but if the density is known (i.e. the weight per unit volume) this can be converted to a volume. If the density of the hydrochloric acid is given as 1.15 g ml^{-1},

$$\text{Density} = \frac{\text{weight}}{\text{volume}}$$

so $$\text{Volume required} = \frac{6.78}{1.15} = 5.89\,\text{ml}$$

To prepare 500 ml of a 0.1 M solution it is therefore necessary to take 5.89 ml of the 27% (w/w) stock and add 494.11 ml of water.

As a final point it is worth noting that in some areas, for example in physical chemistry, concentrations are sometimes given in terms of **molality** (m), not molarity (M). Molality is defined as the number of moles of solute in 1000 g of solvent.

5.3 Accuracy: significant figures and decimal places

If you perform a calculation or take an experimental reading you need to decide what level of accuracy you wish to use. The accuracy is represented by the number of significant figures to which you express the number concerned.

5.3.1 *Significant figures*

If you are asked to express 459 to two significant figures, then only the first two digits can be displayed; all the rest of the digits must be represented by zeros. In this example, since 459 is closer to 460 than to 450, it is logical to round 459 up and say that to two significant figures it equals 460. To express a value in significant figures you therefore look at the last significant figure and decide whether to leave it as it is or increase it by a value of one. In general, if the number after the last significant figure is greater than five, i.e. {6, 7, 8, 9}, than you round up. If the value is less than five, i.e. {1, 2, 3, 4}, then you leave the last significant figure as it is and the number is said to be rounded down. This ensures that you introduce the minimum amount of error possible when rounding numbers off. This is shown in Examples 5.9–11.

Example 5.9

19 732 *to three significant figures:*
The number at position four is less than 5 so we leave the last significant figure as it is. The answer is therefore 19 700.

In Example 5.9 we have therefore 'lost' 32 units, but if we had rounded up we would have added an extra 68 units. Of the two methods, rounding down therefore gives us a value closer to the original number.

Example 5.10

7849 *to two significant figures:*
The number at position three is less than 5, i.e. 7849 is 7800 to two significant figures since it is closer to this value than to 7900.

Example 5.11

379 *to two significant figures:*
The number at position three is greater than 5 so we round up and 379 becomes 380 to two significant figures.

Consider the case where the number after the last significant figure is five. Here you introduce the same error whether you round up or round down (Example 5.12).

Example 5.12

375 *to two significant figures:*
Rounding up to 380 means adding 5 units. Rounding down to 370 means removing 5 units.

In general most people round up in this instance, so 375 would be written as 280 to 2 significant figures. If you are dealing with a large number of values this can increase the level of error you are incorporating into the calculation. It is better to use the rule that if the last significant figure is odd, i.e. {1,3,5,7,9}, then it should be rounded up when the next figure is a five, but if the last significant figure is even it should be rounded down (Example 5.13).

Example 5.13

(a) 375 *to two significant figures:*
     *The number at position three is 5 so you could round up
     or down. The last significant figure (7) is odd, so round
     up.* 375 *becomes* 380 *to two significant figures*

(b) 365 *to two significant figures:*
     *The number at position three is 5 so you could round up
     or down. The last significant figure is even, so round
     down.* 365 *becomes* 360 *to two significant figures.*

If this odd/even rule is applied you should find that your
final answer is more accurate, as shown in Example 5.14:
the use of the odd/even rule means the sum of the three
numbers is 5 units bigger after rounding, compared with a
15-unit increase if all of the numbers are rounded up.

Example 5.14

$$575 + 235 + 325 = 1135$$

*Rounding the figures on the left-hand side to two significant
figures:*

$580 + 240 + 320 = 1140$ *with the odd/even rule*

$580 + 240 + 330 = 1150$ *with the 5-always-rounds-up rule.*

5.3.2 Decimal places

The above examples are all integer values, i.e. whole numbers. Most examples that occur in science contain a fraction
and so can be represented by a decimal. In this case you may
be told to write your answer to a given number of decimal
places. Again, you must decide whether to round the value
up or down, but now you only count digits to the right of the
decimal point, as shown in Example 5.15.

Example 5.15

*Express 2.342 to two decimal places.
The third digit after the decimal point is the value 2, which is
less than 5. 2.342 is therefore 2.34 to two decimal places.*

For decimal places,
count all the figures to
the right of the decimal
point, including zeros

It is important to note that significant figures and decimal
places are not the same, and should not be confused. With
significant figures we only consider the first non-zero values
but with decimal places we count all digits to the right of the

decimal point. This difference is highlighted in Example 5.16.

Example 5.16

(a) *Express 0.0457 to two significant figures.*
   We count only non-zero values so the first figure we count is 4, the second figure is 5, and the third 7. The value at position three is greater than 5; hence to two **significant figures**, *0.0457 is 0.046.*

(b) *Express 0.0457 to two decimal places.*
   The first digit to the right of the decimal point is 0 and the second and third are 4 and 5 respectively. The value at position three is 5, so we round up and 0.0457 becomes 0.05 to two **decimal places**.

Values can be expressed in terms of significant figures and decimal places, but when discussing accuracy you should always use significant figures.

Worked examples 5.3

(a) Represent the following to three significant figures:
   (i) 23.347 893   (ii) 128 904   (iii) 0.003 429
   (iv) 267 491.954

(b) Represent the following to three decimal places:
   (i) 45.096 53   (ii) 0.464 782   (iii) 0.000 89
   (iv) 1 289.632

5.3.3 Accuracy

The level of accuracy shown will depend on two factors.

(1) What level of accuracy is required?

(2) What level of accuracy is possible?

5.3.3.1 The level of accuracy that is required

For example, it is thought that a new drug affects weight gain in humans, so study groups are given varying amounts of the drug and their weight is measured with time. At the start, the weight of one human male is recorded as 75 kg, yet in reality his weight may be 74.536 527 kg. If there really is an effect we would expect to see a reasonable increase in weight. If the weight change is less than 1 kg, then the change would be very small compared with the overall body weight. This small change could be due to a range of factors and may well be a background fluctuation. For the weight change to be meaningful it would therefore have to be greater than 1 kg, so we do not need to be more accurate.

In a second example, suppose you are to weigh out 0.001 g of a chemical. If you choose a 50 g beaker, then you will require scales that measure quite large quantities. The weight of chemical needed is very small compared with the weight of the beaker and so would be almost impossible to measure accurately. You would in this case require a lightweight paper container to allow you to measure this small quantity of material accurately on a set of sensitive scales. It is therefore always necessary to consider what you are measuring and to ensure that the percentage change is measurable.

5.3.3.2 *The level of accuracy that is possible?*

Consider an experiment where we cannot be any more accurate because the scales we are using will not give a more accurate reading. More accurate scales may be costly, and taking the measurements may become difficult with very accurate scales because of fluctuations in the reading.

5.3.3.3 *Accuracy in calculations*

Levels of obesity can be graded using the obesity index, which is calculated from:

$$\text{Obesity index} = \frac{\text{weight (kg)}}{\text{height (m)} \times \text{height (m)}}$$

Suppose the individual mentioned in Section 5.3.3.1 is measured and is 1.72 m tall, then the index would be calculated as:

$$\text{Index} = \frac{75\,\text{kg}}{1.72\,\text{m} \times 1.72\,\text{m}} = 25.351\,54\,\text{kg}\,\text{m}^{-2}$$

Although the calculation is correct, the answer is far too accurate. Your least accurate measurement is the weight, which is measured to two significant figures, so you cannot give the answer to more than two significant figures. Therefore the obesity index = 25 kg m^{-2}. The only exception to this rule is if you take the mean of more than ten data values, in which case it can be given to one significant figure more than the least accurate value (Chapter 11). If you need to perform a calculation, always use the values you have in their most accurate form. For example, we did not say that the height in the above calculation was 1.7 m; we gave it in the most accurate form we could, i.e. 1.72 m. Only at the end of the calculation would you present the answer to the accuracy of your least accurate value.

Worked example 5.4

Evaluate the following, assuming that the values represent experimental data:
(i) 12.354×3.23 (ii) $5 + 4.35 \times 2.3$
(iii) $3.00 \times 2.34 \div 4.001$.

Summary

Concentration refers to an amount in a given volume and can be measured against a number of standards. Concentrations may be given as percentage weight/volume or percentage volume/volume. This is often seen in laboratory manuals when solutions need to be made, since it tells you how much to material to measure out per 100 ml. Another common percent measure is that of percentage weight/weight, which is often used to record the concentration of acids.

The problem with all of the above measurements is that they tell you what quantity of material is present but not its molecular concentration. To express concentration in terms of the number of molecules present, molarity should be used. Whenever a numerical result is expressed, the number of significant figures given should not exceed the number of significant figures in the least accurate measurement involved in the calculation.

End of unit questions

1. A growth medium for *Escherichia coli* contains the components listed in Table 5.1.

Table 5.1

Material	Amount/g litre^{-1}
Glucose	4
Potassium dihydrogenphosphate	10
Magnesium sulphate	0.2
Citric acid	2

(a) What are these concentrations in % (w/v)?

(b) The molecular weights of the above compounds are: glucose $= 180$, potassium dihydrogenphosphate $= 136$, magnesium sulphate $= 120$ and citric acid $= 192$. Calculate the molar concentration of each.

2. How much material would you have to weigh out to prepare 30 ml of tryptone soya broth with the composition described in Table 5.2.

Table 5.2

Material	Amount/% (w/v)
Tryptone	1.7
Peptone	0.3
Glucose	0.25
Sodium chloride	0.5

3. A 25% (w/w) stock of hydrochloric acid has been purchased (specific gravity $1.15\,g\,ml^{-1}$). The molecular weight of hydrochloric acid is 36.5. What volume of the stock is required to make 3 litres of 0.5 M acid?

4. 23.48 g of the buffer N-2-hydroxyethylpiperazine-N'-2-ethanesulphonic acid (HEPES) is weighed out and made up to 180 ml.

(a) What is the concentration in % (w/v)?

(b) If the molecular weight is 238.31, what is the molarity?

5. 34 ml of the amino acid histidine has been prepared as a 0.4 M stock solution. An assay requires 100 ml of reaction mixture containing 0.05% (w/v) histidine. The molecular weight of histidine is 155.2. Can this stock solution be used and if so, how much would you need to make up the 100 ml of solution required?

6. (a) How many grams of sodium hydroxide would be required to make 300 ml of a 0.03 M solution? (Molecular weight of sodium hydroxide = 40.)

(b) What is this as a % (w/v) solution?

7. There is a 5 M stock solution of nitric acid. How many millilitres would be required to make 1500 ml of a 5 mM solution?

8. The density of ethanol is $0.79\,g\,ml^{-1}$ and its molecular weight is 46. How many moles are present in 100 ml?

9. The amino acid glycine has a molecular weight of 75.07 and 1 mg has been dissolved in 1 ml to form a stock solution.

(a) What is the molarity?

(b) My reaction mixture requires a final concentration of 10 μM glycine in 10 ml of buffer. How much stock solution needs to be made up to 10 ml?

6 Tables, Charts and Graphs

6.1 Introduction

One of the functions of a scientist is to gather and interpret data. Furthermore, the interpretation must be presented to other scientists and the data need to support any conclusions made. During the course of an experiment the data are often tabulated. A good scientist will record all the data as they are obtained, showing the correct units and enough detail of the protocol, so that the experiment can be repeated by another worker. In this section we will look at the tabulation of data and frequency tables.

Tabulating data may not be the best way to record them for interpretation or presentation. Data presented in the form of a chart or graph are usually clearer than in a table, since trends may be more apparent. The choice of the correct chart or graph will make an argument clearer but at the same time an inappropriate choice will detract from the argument. The aims of this chapter are:

(a) to introduce frequency tables and the principles behind table construction;

(b) to introduce a range of diagrams and charts to aid data presentation;

(c) to introduce the idea of using figures to show the relationship between two variables.

6.2 Raw data and frequency tables

Data will fall into one of two classes. They measure an outcome either in fixed amounts or over a continuous range. For example, you transform a bacterium with a plasmid (a small extrachromosomal piece of DNA) which carries an antibiotic resistance. To find which bacterial cells have taken up the plasmid you spread the bacteria on an agar plate containing the antibiotic marker so that only those cells with the plasmid can survive. Each plasmid-containing bacterium will give rise to a colony which can be seen by eye. These colonies can be counted to give a measure of the efficiency of the transformation. Each time the experiment is repeated, a different number of colonies will be seen but this number will always be an integer value. You may have eight, twenty or even a hundred colonies but you could not have half values. Since the results are measured in fixed amounts, these are termed **discrete data**.

In some cases the data can take any value over a range; for example, the height of a house plant called *Saintpaulia ionantha* (African violet) ranges up to 10 cm. The height could take any value over this range and hence the height measurement would be an example of a **continuous data measurement**. Both discrete and continuous data can be tabulated.

6.2.1 Table preparation

A clear way to present data in a laboratory book is by the use of tables. The table should contain clearly labelled columns, each of which should show units where appropriate. An example is given in Table 6.1 for the case of a protein assay. Five identical samples were assayed for the presence of protein using the Bradford method, in which a blue dye interacts with the protein and the amount of colour is measured using a spectrophotometer.

Table 6.1 Data from a Bradford assay

Observation no.	Protein concentration/mg ml^{-1}
1	1.00
2	0.98
3	1.12
4	1.08
5	0.99

Source: Data from classwork at University of Central Lancashire.

Notice that the units of protein concentration are clearly given and separated from the text by a solidus, as described in Chapter 3. The solidus does not mean 'per'. In some texts the units are placed in brackets. The column heading would then read:

Protein concentration (mg ml^{-1})

As we will see later in this chapter, displaying data graphically allows the audience to observe general trends, but sometimes it is desirable to display the actual data. In this case tables are ideal since they allow quantitative features to be observed. A clear table will allow obvious patterns to be

Tables allow quantitative features to be displayed

discerned but usually the trends will have to be described in the text. There are some simple rules which, if followed, make tables easier to interpret.

1. The table needs to be able to convey information: therefore, keep it simple. If there are many data, consider whether it would be better to split them and have more than one table.

2. You should be clear about what you intend to show in the table. Once the table's purpose has been determined, the main data headings should be arranged horizontally so that the data being compared are arranged in columns. It has been shown that people find it easier to compare columns of data rather than rows. Example 6.1 shows the same data arranged in different ways in Tables 6.2 and 6.3.

Example 6.1

The route by which a drug is administered can affect its toxicity. The toxicity of three agents against mice is recorded below. Toxicity is measured in terms of the LD_{50}, i.e. the amount of agent required to kill 50% of the population.

Table 6.2

	Oral	Intramuscular	Intravenous	Subcutaneous
Pentobarbital	279.59	124.27	79.76	130.42
Isoniazid	142.23	139.95	153.30	160.19
Procaine	495.76	645.23	46.78	804.23

It would appear that the function of this table is to compare the toxicity of the drugs when they are applied via different routes, i.e. we want to show how the toxicity of pentobarbital, for example, varies when it is taken by the four routes listed. We are not comparing the three drugs. The data should therefore have been arranged as shown in Table 6.3.

Table 6.3

	Pentobarbital	Isoniazid	Procaine
Oral	279.59	142.23	495.76
Intramusuclar	124.27	139.95	645.23
Intravenous	79.76	153.30	46.78
Subcutaneous	130.42	160.19	804.23

3. All tables should have a comprehensive explanatory title. Columns and rows should be clearly labelled and, where appropriate, units should be shown. There should

also be a clear indication of the source of the data. Taking these points into consideration, Table 6.3 needs to be modified to the form shown in Table 6.4.

Table 6.4 Table to show how toxicity varies with the route of drug administration

Route of administration	Pentobarbital LD_{50}*/mg kg^{-1}	Isoniazid LD_{50}*/mg kg^{-1}	Procaine LD_{50}*/mg kg^{-1}
Oral	279.59	142.23	495.76
Intramuscular	124.26	139.95	645.23
Intravenous	79.76	153.30	46.78
Subcutaneous	130.42	160.19	804.23

Source: The data have been based on values from T. A. Loomis (1968) *Essentials of Toxicology*, Philadelphia: Lea and Febiger.
*Mouse toxicity data.

4. Notice that in Table 6.4 there was a lot of information regarding the toxicity data. The columns could have been headed:

    'Pentobarbital LD_{50}/mg kg^{-1} based on mouse toxicity data.'

    This is obviously far too lengthy and would detract from the table so extra detail regarding the species was placed in a footnote. Footnotes are commonly used and can include information regarding changes, omissions, approximations, experimental detail and any other factors which are felt to be essential.

5. The data should be recorded as accurately as possible (Chapter 5). However, if the table is to form part of a report it is recommended that entries are limited to two or three significant figures since the greater the number of digits present, the more difficult it is for the reader to digest the information. We can use this to modify Table 6.4. Notice that in Table 6.5 a note on the accuracy has been included in the footnote. The clarity would be even greater if only two significant figures were used; but it should be remembered that rounding figures in this way loses accuracy, and one of the functions of tables is to make a quantitative presentation of the data.

6. The use of lines and different text styles can help make tables more striking, and at the same time even easier to follow. If the table contains a number of columns with common headings then clarity can be further increased by the use of a single heading. For example in Table 6.5 there are three columns listing toxicity data so the table can be transformed as shown in Table 6.6.

Table 6.5 Table to show how toxicity varies with the route of drug administration

Route of administration	Pentobarbital LD_{50}*/mg kg^{-1}	Isoniazid LD_{50}*/mg kg^{-1}	Procaine LD_{50}*/mg kg^{-1}
Oral	280	142	496
Intramuscular	124	140	645
Intravenous	79.8	153	46.8
Subcutaneous	130	160	804

Source: The data has been based on values from T. A. Loomis (1968), *Essentials of Toxicology*, Philadelphia: Lea and Febiger.
*Mouse toxicity data displayed to three significant figures.

Table 6.6 Table to show how toxicity varies with the route of drug administration

Route of administration	Toxicity LD_{50}*/mg kg^{-1}		
	Pentobarbital	Isoniazid	Procaine
Oral	280	142	496
Intramuscular	124	140	645
Intravenous	79.8	153	46.8
Subcutaneous	130	160	804

Source: The data has been based on values from T. A. Loomis (1968), *Essentials of Toxicology*, Philadelphia: Lea and Febiger.
*Mouse toxicity data displayed to three significant figures.

7. Although not possible in the case of Table 6.6, it is also useful if the columns of data can be arranged in order of size, since this makes it easier to observe trends. In addition percentages and ratios can at times be used to clarify points within the table. For example, suppose that the point of interest was how the route of drug administration affects pentobarbital toxicity, then we may be better arranging the data as in Table 6.7. Notice though that by only including relative toxicities some information has been lost since the original data points have not been shown.

Table 6.7 Table to show how the toxicity of pentobarbital varies with the route of administration

Route of administration	Relative toxicities
Oral	1.00
Subcutaneous	0.46
Intramuscular	0.44
Intravenous	0.29

Source: Relative toxicities are based on data from T. A. Loomis (1968), *Essentials of Toxicology*, Philadelphia: Lea and Febiger.

6.2.2 *Frequency tables*

Tables can be useful for recording the frequency with which a given result occurs. For example, swabs were taken from 30 surgical wound infections and tested for the presence of a range of bacteria. The occurrence of bacteria in the wounds is recorded in Example 6.2 using a **tally chart** (Table 6.8). This is simply a table which contains a tally. Each time an observation is made, a single stroke is entered by that result in the table.

Example 6.2

Table 6.8 Frequency with which the listed bacteria were found to have colonised infected surgical wounds

Micro-organism	Tally
Staphylococcus aureus	IIII III
Escherichia coli	IIII I
Pseudomonas aeruginosa	III
Proteus spp.	III

Source: Data are fictitious but based on known examples of wound colonisation.

Notice that tally marks are grouped in sets of five to make it easier to add up the number of times a result has occurred. The fifth occurrence in each set of five marks is recorded by placing a line diagonally through the other four tally marks as shown in Table 6.8. The number of occurrences is termed the **frequency** and it is usual to prepare a **frequency table** for this type of data. Frequency tables list the frequency of occurrence for each possible result. The table is said to show the **frequency distribution**; an example is given in Table 6.9.

It may be that you are interested in how often a result was obtained, compared with all the observations recorded. This is termed the **relative frequency** and is obtained by dividing the frequency of occurrence by the total number of occurrences. Table 6.10 shows the **relative frequency distribution**

Table 6.9 Frequency with which the listed bacteria were found to have colonised infected surgical wounds

Micro-organism	Frequency
Staphylococcus aureus	8
Escherichia coli	6
Pseudomonas aeruginosa	3
Proteus spp.	3
TOTAL	20

Source: Data are fictitious but based on known examples of wound colonisation.

for Example 6.2. Notice that at the bottom of the columns we record the total number of occurrences. In addition we can multiply the relative frequency by 100 to give it as a percentage of the total (Chapter 2).

Table 6.10 Frequency with which the listed bacteria were found to have colonised infected surgical wounds

Micro-organism	Frequency	Relative frequency	Percentage
Staphylococcus aureus	8	0.40	40
Escherichia coli	6	0.30	30
Pseudomonas aeruginosa	3	0.15	15
Proteus spp.	3	0.15	15
TOTAL	20	1.00	100

Source: Data are fictitious but based on known examples of wound colonisation.

Suppose you were recording a continuous variable or had taken a large number of readings. It becomes cumbersome to use a frequency table of the kind shown in Table 6.10, and the data can be represented more clearly in a table which contains ranges or groups. These groups are termed **classes** and the table is said to show a **grouped frequency distribution.** This form of table allows data to be shown in a concise and clear manner but you need to realise that some information is lost since you no longer know the exact values of the results.

Grouping data leads to the loss of information

Example 6.3

The weight of 30 babies was recorded at birth in kilograms. The results are recorded in Table 6.11.

Table 6.11 Birth weight of babies born in the UK after 38–42 weeks' gestation

Birth weight/kg	Frequency
2.5–2.9	4
3.0–3.4	7
3.5–3.9	11
4.0–4.4	8
TOTAL	30

Source: Data are fictitious but based on the weight distribution of newborn babies as given in the centile charts supplied by the Health Education Authority (1995). Reproduced with kind permission from the Health Education Authority.

The values which show the range for each class are termed the **class limits**. In the first row of Table 6.11 the **lower class limit** is given by 2.5 kg and the **upper class limit** by 2.9 kg. Notice, though, that although the first class ends at 2.9 kg, the second class does not start until 3.0 kg. The first class must therefore range from 2.45 kg up to but not including 2.95 kg. The second class starts at 2.95 kg and ranges up to but not including 3.45 kg. These values are termed the **class boundaries** and are obtained by adding the upper class limit of one group to the lower class limit of the next group and dividing by two.

Example 6.4

The boundary between the first and second classes (Table 6.11) was therefore obtained as follows:

$$(2.9 + 3.0) \div 2 = 2.95 \, \text{kg}$$

The **class width** is given by finding the difference between the upper and lower boundaries.

Example 6.5

In Table 6.11 all the classes have the same class width, for example:

$$2.95 - 2.45 = 0.5 \, \text{kg}$$

Notice from Example 6.5 that, using the class boundaries, the class width is found to be 0.5 kg, not 0.4 kg as would appear from the class limits.

Since we no longer know the exact values for the data points if an estimate is needed, the **class midpoint** should be used. This is obtained by taking half the class width and adding it to the lower class boundary, as shown in Example 6.6.

Example 6.6

From Table 6.11,

Class width $= 0.5 \, \text{kg}$

Lower class boundary for first class $= 2.45 \, \text{kg}$

Class midpoint $= 2.45 + (0.5/2) = 2.70 \, \text{kg}$

The class midpoints are used in Chapter 11 to find arithmetic means from grouped frequency tables.

The final point to note is that in some cases the table may record the **cumulative frequency**. This is simply obtained by adding each successive frequency and it acts as a running total, showing how many values have been obtained as you move from the top to the bottom row of the table. The final value in the cumulative frequency column should equal the sum of all the frequencies (Table 6.12).

Table 6.12 Birth weight of babies born in the UK after 38–42 weeks' gestation

Birth weight/kg	Frequency	Cumulative frequency
2.5–2.9	4	4
3.0–3.4	7	11
3.5–3.9	11	22
4.0–4.4	8	30
TOTAL	30	30

Source: Data are fictitious but based on the weight distribution of newborn babies as given in the centile charts supplied by the Health Education Authority (1995). Reproduced with kind permission from the Health Education Authority.

The cumulative frequency is useful because at a glance it allows you to see the number of occurrence up to and including any classes of interest. For example, we know from Table 6.12 that 22 of the babies had a weight up to but not including 3.95 kg.

6.3 Charts, diagrams and plots

Unless it is essential that you present all the data values for quantitative analysis, then data are often better presented in the form of a figure. Figures are more visual than tables and can attract and hold the readers' attention more easily. The higher visual impact can highlight patterns, making trends more obvious. The main disadvantage of figures is that you lose accuracy since it is usually impossible to read the exact value of the data point from the figure. The choice of figure will depend on what you are trying to show and the nature of your data. Some commonly used methods of displaying data are described below, but in all cases the figure should include a clear title and a note regarding the source of the data.

Displaying data in the form of figures means some detail is lost

6.3.1 *Pictograms*

Pictograms simply involve displaying data by using a symbol to represent the quantity measured. The number of symbols drawn therefore represents the amount of material in that category. This is a simple means of presentation and its main advantages are that it can be eye-catching and simple trends can be observed rapidly.

Example 6.7

The standardised mortality rates (SMRs) for coronary heart disease (CHD) represent the number of deaths due to CHD per 100 000 of the population. Some data are given in Table 6.13. relating to male mortality in the period 1986–1988.

Table 6.13 SMRs for CHD

Country	SMRs for CHD
England and Wales	200
Northern Ireland	275
Japan	25
USA	175

Source: Adapted from Ashwell (1993), *Diet and Heart Disease*. The British Nutrition Foundation. Reproduced with kind permission from The British Nutrition Foundation.

The data can be represented as the pictogram in Figure 6.1.

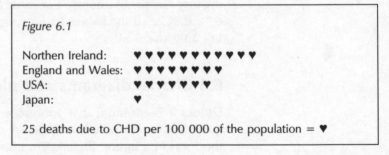

Figure 6.1

Northen Ireland: ♥ ♥ ♥ ♥ ♥ ♥ ♥ ♥ ♥ ♥ ♥
England and Wales: ♥ ♥ ♥ ♥ ♥ ♥ ♥ ♥
USA: ♥ ♥ ♥ ♥ ♥ ♥ ♥
Japan: ♥

25 deaths due to CHD per 100 000 of the population = ♥

Pictograms can include fractions of a symbol, so in this example 62.5 deaths per 100 000 of the population would have been represented by two and a half hearts.

6.3.2 Pie charts

A pie chart is used to show relative frequencies, i.e. it represents parts of a whole. It is drawn as a circle which represents the 100% value and it is divided into sections, each of which represents a given category. The size of each section depends on what percentage of the whole that category represents. A circle contains 360°, so to find the angle for each slice of the circle you multiply the relative frequency by 360.

Pie charts are used to display relative frequencies

In Table 6.10 we calculated the relative frequencies of occurrence for four bacteria in surgical wound infections. The relative frequency for the *Proteus* spp. was 0.15, so the angle subtended at the vertex of this 'piece of pie' would be $0.15 \times 360 = 54°$ (Figure 6.2).

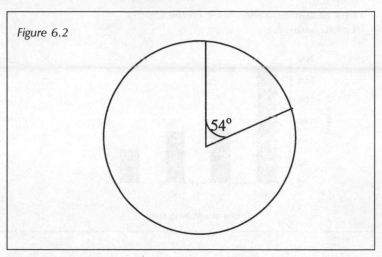

Figure 6.2

Repeating this for the three remaining bacteria we obtain Figure 6.3.

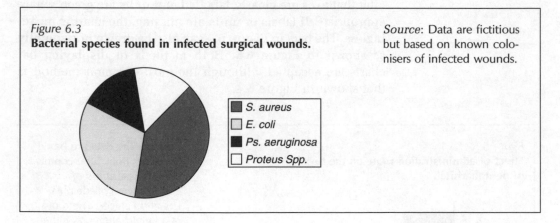

Figure 6.3
Bacterial species found in infected surgical wounds.

S. aureus
E. coli
Ps. aeruginosa
Proteus Spp.

Source: Data are fictitious but based on known colonisers of infected wounds.

6.3.3   *Bar charts*

The frequency with which a result occurs can be represented by drawing a rectangular box or **bar**. The height of the bar corresponds to the frequency of occurrence, but all the widths should be the same to avoid confusion. This form of chart is ideal for displaying discrete data sets and the bars are separated from each other by spaces. The vertical and horizontal lines which define the dimensions of the bar chart are termed **axes**. The axis used to measure bar height

Bar charts are ideal for displaying discrete data sets

will have a clearly labelled scale but the other one simply contains labels denoting the data sets. This is illustrated in Figure 6.4, using the data from Table 6.6.

Figure 6.4
Effect of administration route on the toxicity of pentobarbital.

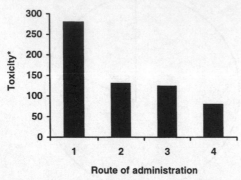

Route of administration

Source: The data are based on values from T. A. Loomis (1968), *Essentials of Toxicology*. Philadelphia: Lea and Febiger. The route of administration was either oral (1); subcutaneous (2); intramuscular (3); or intravenous (4).

*Mouse toxicity data represented as $LD_{50}/mg\,kg^{-1}$.

Notice that each bar is clearly labelled but since the labels are quite long, details are placed in the figure legend. It is vital that axes are clearly labelled and units are given where appropriate. If labels or units are missing, the chart is meaningless. The bars in Figure 6.4 can also be drawn horizontally as shown in Figure 6.5. Both methods of displaying bar charts are accepted, although the most common method is that shown in Figure 6.4.

Figure 6.5
Effect of administration route on the toxicity of pentobarbital.

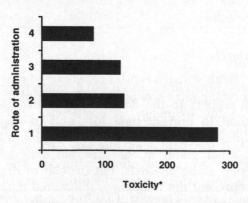

Toxicity*

Source: The data are based on values from T. A. Loomis (1968), *Essentials of Toxicology*. Philadelphia: Lea and Febiger. The route of administration was either oral (1); subcutaneous (2); intramuscular (3); or intravenous (4).

*Mouse toxicity data represented as $LD_{50}/mg\,kg^{-1}$.

The length of the bar is proportional to the magnitude of the data

The key point to remember is that all the bars in bar charts have the same width and the frequency of occurrence is given by the length of the bars. Bar charts can be useful for allowing comparison of data sets, and sets of data can be grouped and distinguished by colour or shading. For example, the toxicity data for all three drugs shown in Table 6.6 can be compared (Figure 6.6).

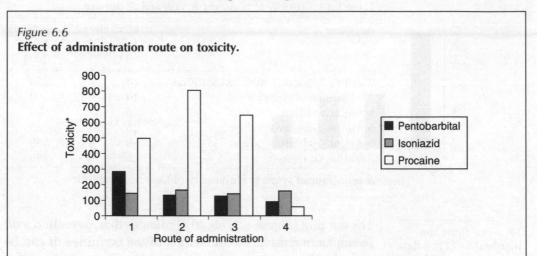

Figure 6.6
Effect of administration route on toxicity.

Source: The data are based on values from T. A. Loomis (1968), *Essentials of Toxicology*. Philadelphia: Lea and Febiger. The route of administration was either oral (1); subcutaneous (2); intramuscular (3); or intravenous (4). *Mouse toxicity data represented as $LD_{50}/mg\,kg^{-1}$.

Modern graphics packages will allow the bars to be represented in either the standard form shown in Figure 6.6 or a three-dimensional form. Bars can even be shown next to each other or arranged behind each other. When using these features, you should remember that it is no use producing a great-looking chart if the trends are no longer easy to observe. When presenting data the most useful rule is to keep things simple: show the trend or relationship as clearly and concisely as possible.

6.3.4 *Dot plots*

In the bar charts shown in Figures 6.4–6.6, much information was lost with respect to the data values themselves. The simplest method to prevent data loss and to display as much information as possible is to use a dot plot. It is comparable with a bar graph in that discrete data points or classes are listed along one axis, but instead of drawing a rectangle to represent the frequency, the actual data points themselves are listed. This is shown in Example 6.9.

Example 6.9

A range of acridines were tested for their antibacterial effect against Staphylococcus aureus *and* Escherichia coli. *The data are given in Table 6.14 with respect to the minimum concentration required to kill the cells (minimum lethal concentration, MLC).*

Table 6.14   Toxicity tests against *E. coli* and *S. aureus*

Acridine	MLC vs S. aureus (µM)	MLC vs E. coli (µM)
9-amino-3-chloro-7-methoxyacridine	10	250
3,9-Diamino-7-ethoxyacridine	10	100
9-Aminoacridine	50	50
9-Amino-3-methylacridine	25	25
9-Amino-3-chloroacridine	50	100
Acridine Orange	25	250

Source: Project work at University of Central Lancashire.

Dot plots show the distribution of the data points and identify any outliers

The dot plot (Figure 6.7) clearly indicates that, overall, *E. coli* seems more resistant to the action of the acridines. It can be seen that this plot preserves the data values and it has the advantage that any points which do not seem representative of the sample can be identified and investigated. These non-representative points are termed **outliers** (Chapter 11). Dot

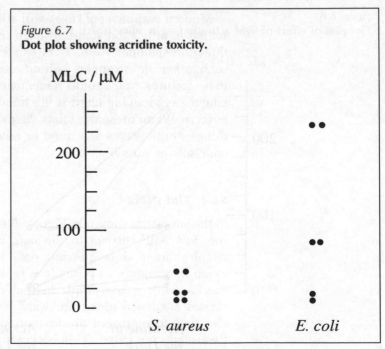

Figure 6.7
Dot plot showing acridine toxicity.

plots can be very useful for observing the data distribution, but if there are many data points they can be tedious to prepare.

One major advantage with dot plots is that they can be used effectively to link sets of observations.

Example 6.10

The acridines listed in Table 6.14 can absorb light and when they do so they produce a highly toxic chemical called a free radical. The effect of illumination on acridine toxicity was therefore observed and the data for E coli are shown in Table 6.15, and compared to dark toxicity by linking the dots for each acridine on the dot plot in Figure 6.8.

Table 6.15   Light activation of acridines

Acridine	MLC vs E. coli in the dark (µM)	MLC vs E. coli in the light (µM)
9-Amino-3-chloro-7-methoxyacridine	250	100
3,9-Diamino-7-ethoxyacridine	100	10
9-Aminoacridine	50	50
9-Amino-3-methylacridine	25	25
9-Amino-3-chloroacridine	100	100
Acridine Orange	250	25

Source: Project work at University of Central Lancashire.

Figure 6.8
Dot plot of effect of light activation on acridine toxicity.

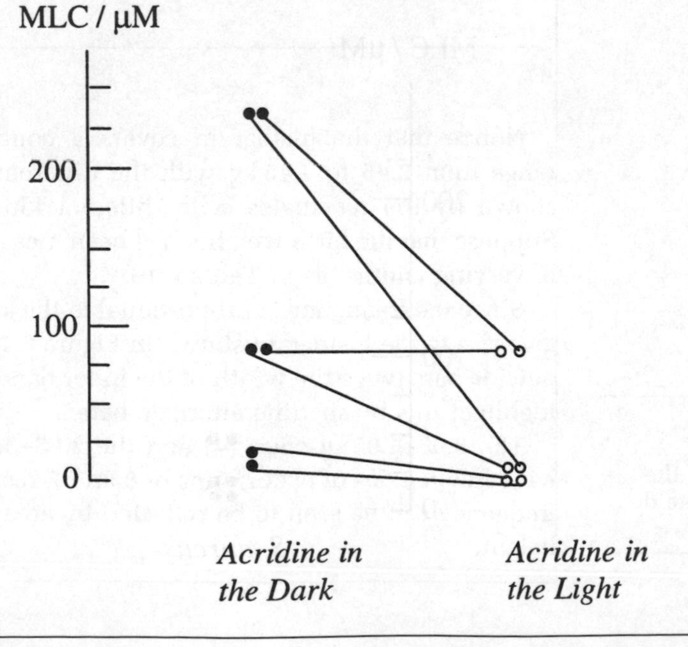

6.3.5 *Histograms*

Histograms are useful
for displaying
continuous or grouped
data

A histogram is similar to a bar chart but the two should not be
confused. A bar chart is ideal for representing discrete data
since each bar represents one value. A histogram is useful for
representing continuous or grouped data. Because the data
are continuous, it is the area of the rectangle which relates
to the frequency, not its height. Class widths are represented
by the width of the rectangle. If all the class widths are the
same, then the widths of the bars are the same so the frequency is proportional to the height as with the bar chart.

The widths of the
rectangles correspond
to the class boundaries

When the rectangles are being drawn, the widths must correspond to the class boundaries and not to the class limits. The
birth weights shown in Table 6.12 are illustrated in Figure 6.9
using a histogram.

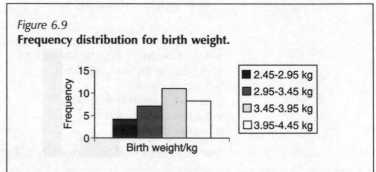

Figure 6.9
Frequency distribution for birth weight.

Source: Data are fictitious but based on the weight distribution of
newborn babies as given in the centile charts supplied by the
Health Education Authority (1995). Reproduced with kind permission from the Health Education Authority.

Notice that the histogram covers a continuous weight
range from 2.45 to 4.45 kg with the different classes being
shown by the rectangles with different kinds of shading.
Suppose that the birth weights had been measured in classes
of varying widths, as in Table 6.16.

Since the frequency is proportional to the area, this would
give rise to the histogram shown in Figure 6.10 with the two
outside bars twice the width of the inner bars. I have tried to
highlight this by shading alternate bars.

The area of the
rectangle is
proportional to the
magnitude of the data
value

The 3.95–4.95 kg class (4) and the 2.95–3.45 kg class (2)
have frequencies of occurrence of 8 and 7 respectively. This
frequency can be seen to be reflected by area but not by the
height.

Table 6.16 Birth weight of babies born in the UK after 38–42 weeks' gestation

Birth weight/kg	Frequency	Cumulative frequency
2.0–2.9	4	4
3.0–3.4	7	11
3.5–3.9	11	22
4.0–4.9	8	30
TOTAL	30	30

Source: Data are fictitious but based on the weight distribution of newborn babies as given in the centile charts supplied by the Health Education Authority (1995). Reproduced with kind permission from the Health Education Authority.

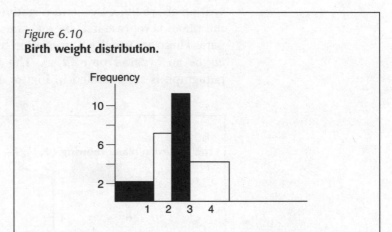

Figure 6.10
Birth weight distribution.

The classes represent 1.95–2.95 kg (1), 2.95–3.45 kg (2), 3.45–3.95 kg (3) and 3.95–4.95 kg (4). Reproduced with kind permission from the Health Education Authority.

6.3.6 *Scatter graphs*

A scatter graph is a graphical way of representing what is termed a **function**. It shows the relationship between two variables x and y. The horizontal axis runs from left to right and is termed the **x-axis** or **abscissa**. The vertical axis runs from the bottom of the page to the top and is termed the **y-axis** or **ordinate**. These two axes intercept at the **origin** and the area which they describe is termed the **x–y plane** or the **Cartesian plane**. When a scatter graph is being plotted, the x variable; should always be the independent variable; this is usually the element you are controlling in the experiment and it does not depend on the other variable. The y variable is the dependent variable, i.e. it depends on the value of x

---

A scatter graph
displays the function
linking the dependent
and independent
variables

---

and it is usually the variable you are measuring. For example, you are measuring the amount of product produced by an enzyme catalysed reaction with respect to time. You take measurements at 1, 2, 5 and 10 minutes. The independent variable is time, since time does not depend on the amount of product obtained. This is what you are controlling, and it forms the x-axis. The amount of product formed is dependent on time; it is therefore the dependent variable and this is what you are measuring. This forms the y-axis. To be able to use the graph, both axes must have a clearly marked scale. The origin is taken to be represented by $x = 0$ and $y = 0$. Positive x values are written to the right of the y-axis and negative values to the left. Similarly, positive y values are written above the x-axis and negative below it. Each point in the plane is represented by an x coordinate and a y coordinate. These are usually represented by writing them in brackets as an **ordered pair** (x, y). The information in the last paragraph is summarised in Figure 6.11.

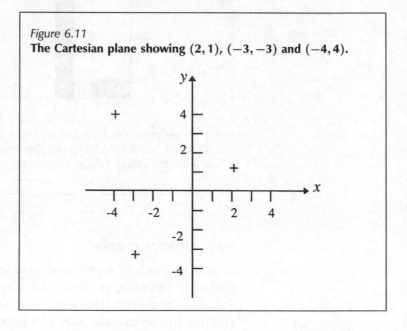

Figure 6.11
The Cartesian plane showing $(2, 1)$, $(-3, -3)$ and $(-4, 4)$.

Many computer packages can accurately place a range of symbols in the plane to allow you to distinguish between different points; but if you are drawing the graph by hand, the points represented by the (x, y) coordinates should be marked on the plane using a sharp pencil by placing a cross at the correct position. The centre of the cross represents the point. The use of dots, circles etc. can make it hard to distinguish exactly which point the symbol represents. If the points lie on a straight line you can use a ruler to draw a line of **best fit**. This involves balancing the points so that there is an equal number on both sides of the line. The sum of the distances from each of the points above the line to the line should equal the sum of the distances from each of the points below the line to the line. Statistical packages will do this for you using, for example, **linear regression**. If the points do not fall on a straight line they should be fitted with a smooth curve. You should not simply join up the points one at a time since it is likely that some of the measurements are inaccurate or incorrect; by using a line of best fit you are 'averaging out' this error. An example of a line of best fit, done by eye, is given in Figure 6.12.

Example 6.11

Known concentrations of protein were assayed using the Bradford assay. This assay involves adding a dye to the protein solution. The dye interacts with the protein and the solution becomes blue. The more coloured the solution, the more protein is present. The level of colour can be detected by measuring the absorption of light at the appropriate wavelength.

Table 6.17 Bradford assay

Protein concentration/μg ml^{-1}	Absorption reading at 595 nm
0	0.00
20	0.12
40	0.21
60	0.39
100	0.60
120	0.74

Source: Based on student data obtained at the University of Central Lancashire.

Avoid extrapolation since the relationship between the variables may change after the last plotted point

Notice that in Figure 6.12 the line ends at the last point so it does not continue past the value given by the last piece of data. Extending the line without data is termed **extrapolation** and should be avoided since you have no experimental evidence to show that the relationship between the absorption and the protein concentration continues in a straight line past that last point. At higher concentrations the protein might aggregate and therefore precipitate, so the graph could start to curve. Calibration curves are widely used in life sciences: known quantities are measured to give a standard curve, and this is used to determine an unknown.

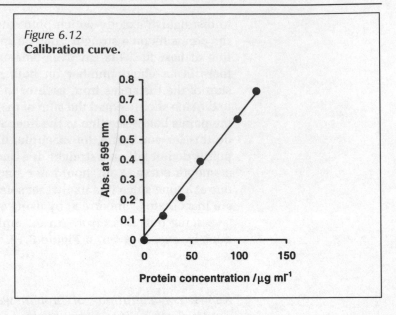

Figure 6.12
Calibration curve.

Suppose, for example, the calibration curve in Figure 6.12 had been prepared using serum protein from blood. A sample of blood has been taken from a patient and the blood serum protein measured using the Bradford assay. The absorption was found to be 0.54 absorption units at 595 nm. If you find this value on the y-axis of the plot, then read across to the calibration curve and drop down to the x-axis, this gives you the concentration of protein in your unknown sample. This is illustrated in Figure 6.13 where the unknown is found to be approximately $90 \, \mu\text{g ml}^{-1}$.

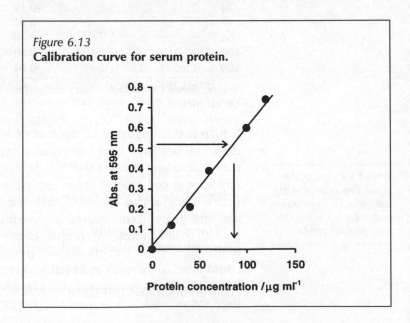

Figure 6.13
Calibration curve for serum protein.

The final point to note is one regarding practicality. The graphs should be large enough so that they are clear to read. The scales should be chosen so that the points are easy to plot and so that readers can easily understand them. On some occasions you may not require the scale to start at zero. For example, it can be seen in Figure 6.14 that the origin is still at $(0, 0)$ but the zig-zag line on the x-axis indicates that the scale is not linear before the value 10, i.e. the distance from zero to ten has been condensed. The graph would then be plotted as normal; the point $(11, 2)$ is shown.

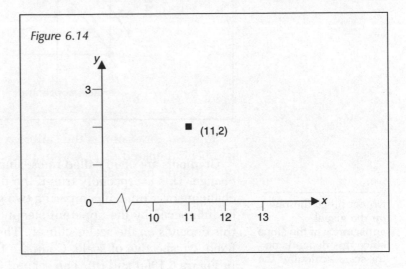

Figure 6.14

6.3.6.1 *Intercepts and gradients*

Line graphs can be characterised in a number of ways, but two parameters of importance are the intercept and the gradient. The point at which the line crosses the axis is termed the **intercept**. In Figure 6.15 the line crosses the y-axis at the point where $y = c$ and this is said to be the y intercept. The slope of the line is termed the **gradient** and is given by the equation

$$\text{gradient} = \frac{y_1 - y_0}{x_1 - x_0}$$

where there are two points (x_0, y_0) and (x_1, y_1) such that $x_0 < x_1$.

The change in x and y is often denoted by capital delta (Δ), and many textbooks will simply say:

$$\text{gradient} = \Delta y / \Delta x$$

This is illustrated in Figure 6.15.

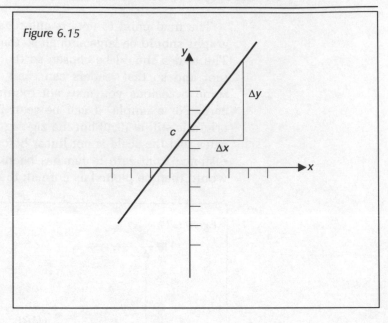

Figure 6.15

Gradients are often called rates, since they give the rate of change. This is especially true if the plot shows a variable y changing over time. If comparing two graphs you should not be influenced by the apparent steepness of the slope, since this depends on the scales chosen. The gradient will remain fixed, irrespective of scale. Compare the two graphs shown in Figure 6.16(a) and (b). The second graph looks as though the rate is much faster because the lines are steeper but both (a) and (b) have the same gradients; in fact they show the same data but plotted on different scales.

Do not judge the rate on the visual appearance of the slope since this depends on the scale—calculate the gradient

6.3.6.2 Gradients of curves

If the graph contains a curve, then the gradient obviously changes as you move along the curve. Consider the graph in Figure 6.17.

As you move from −5 to zero on the x-axis the graph slopes downwards and the slope changes. For example, between −5 and −4 the slope is very steep but between −1 and 0 it is much more gradual. As you move from 0 to 5 on the x-axis the slope changes direction. To find the gradient of a curve, the best method is to find the equation and use calculus to determine the slope, but this is beyond the scope of this text. A quick alternative which allows you to find the gradient at any given point is to draw a straight line which just touches the curve at the point of interest and

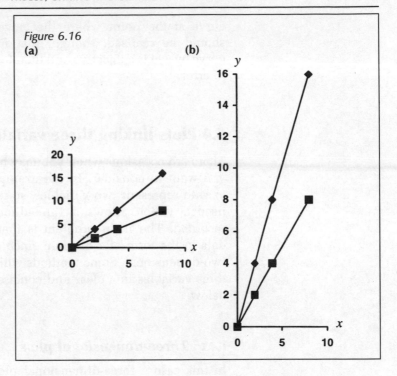

Figure 6.16

(a)   (b)

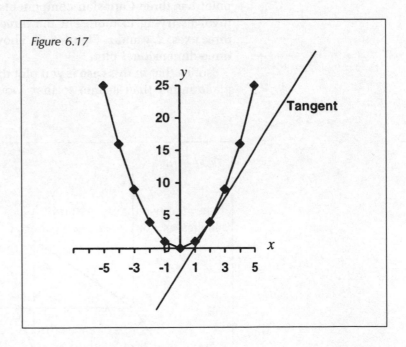

Figure 6.17

which seems to follow the slope of the curve at this point
(Figure 6.17). This line is termed a **tangent**. The gradient of
the tangent can be taken to approximate the gradient of the

curve at the point where the tangent and curve meet. It should be realised, though, that fitting a tangent to the curve by eye is subjective and therefore prone to some degree of error.

6.4 Plots linking three variables

There are occasions when you may have three variables that you want to consider. In the example above we used two axes to represent two variables, so in this case a method is needed which allows a three-dimensional system to be recorded. The main problem is therefore representing the data in the form of a diagram since diagrams are obviously two-dimensional. Some methods which attempt to represent three variables in a clear and concise manner are described below.

6.4.1 *Three-dimensional plots*

In this case a three-dimensional plot is made where each point has three Cartesian components (x, y, z). This method involves trying to represent the three variables by drawing three axes, x, y and z. Figure 6.18. shows point P plotted on a three-dimensional plot.

Notice that in this case if you plot three points they lie in a plane rather than a line, so they would be connected by a

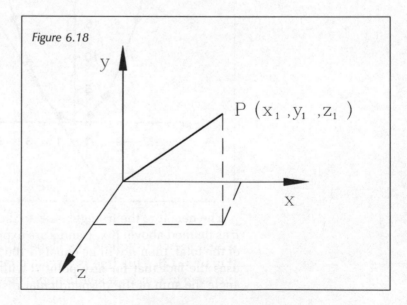

Figure 6.18

surface. These plots can be hard to decipher and even harder to draw.

6.4.2 *Triangular charts*

This is a simple method of representing three variables, but it only uses a two-dimensional plot rather than attempting to represent a three-dimensional plot as in Section 6.4.1. A triangular plot will only function if the three variables are linked in such a way that when added they always give the same constant:

$$x + y + z = h$$

where x, y and z are variables and h is a constant. This is ideal if you are studying three components which together make up the whole, since each factor can be represented as a percentage of the total, i.e.

$$x + y + z = h = 100\%$$

A triangular plot involves drawing an equilateral triangle, i.e. a triangle in which all three sides are the same length and the angles subtended are 60° (Figure 6.19).

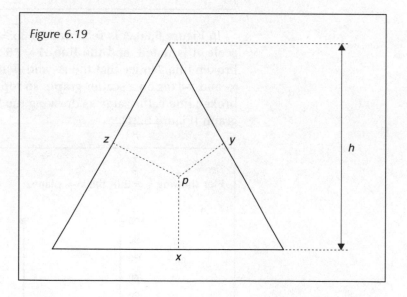

Figure 6.19

The height of the triangle is set to the value of the constant h as defined above. If x, y and z are expressed as a percentage of the total, then h will be equal to 100%. The triangular plot uses the fact that for any point P within the equilateral triangle the sum of the perpendicular distances between P and the sides of the triangle is equal to the height of the triangle. If each side of the triangle is given a scale of 0–100, to allow the representation of the variables we can plot (x, y, z) as percentages (Figure 6.20).

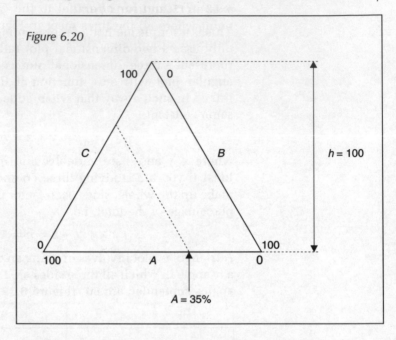

Figure 6.20

In Figure 6.20 A is recorded as 35% so the 35% mark on scale A is noted, and the line $A = 35$ is represented by the broken line. Notice that the A- and B-axes are acting like the x- and y-axes on a scatter graph, so representing $A = 35$ by a broken line is the same as drawing the line $x = 2$ on a scatter graph (Figure 6.21).

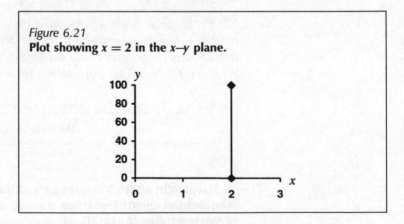

Figure 6.21
Plot showing $x = 2$ in the x–y plane.

In Figure 6.21 the line $x = 2$ crosses the x-axis at 2 and runs parallel to the y-axis (i.e. $x = 0$). On a triangular chart

(Figure 6.20) the same principles apply but in this case the axes are at 60° rather than at right angles. For example, the line at $A = 35$ crosses the A-axis and runs parallel to the B-axis (i.e. $A = 0$). Similarly the line at $B = 15$ crosses the B-axis at 15 and runs parallel to the C-axis (i.e. $B = 0$). The point where all the lines meet in Figure 6.22 is unique and represents the three values $A = 35$, $B = 15$ and $C = 50$, i.e. $(A, B, C) = (35, 15, 50)$.

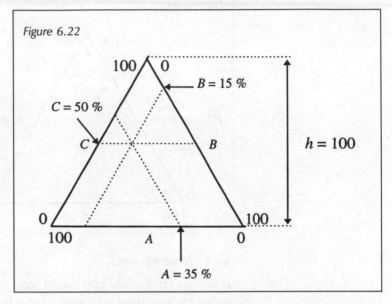

Figure 6.22

This plot can prove very useful for identifying relationships between variables, as illustrated by Example 6.12.

Example 6.12

Three bacteria were grown in a mixed culture and treated with a range of potential antibacterials. Each bacterial population was recorded (Table 6.18) in terms of its contribution to the total bacterial population after drug treatment.

Table 6.18 Effects of drug treatment on a mixed culture

Drug	Bacteria remaining after treatment (%)		
	Escherichia coli	*Pseudomonas aeruginosa*	*Bacillus cereus*
Proflavin (0.01 mM)	24	70	6
Acridine Orange (0.01 mM)	28	72	10
9-Aminoacridine (0.01 mM)	5	90	5

Source: Theoretical data based on MLCs observed at University of Central Lancashire.

It can be seen from Figure 6.23. that one of the compounds seems to almost totally prevent the growth of *E. coli* and *B. cereus* in comparison with *Ps. aeruginosa*. Remember, though, that these are relative values; all three bacteria have probably suffered growth inhibition, but this plot shows the effect of the agent on each strain relative to the other strains.

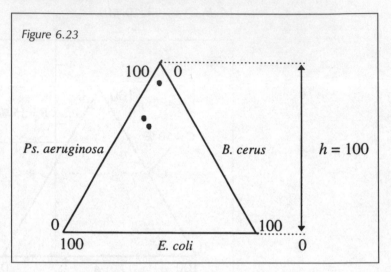

Figure 6.23

6.4.3 Nomograms

If you go for a walk, you could map your route on a piece of paper by recording where you have been. Now suppose your walk was in a mountainous region. On the paper you may have travelled from a point (x_1, y_1) to a point (x_2, y_2) and covered 1 km, but during this move you may have climbed up the face of a mountain! Your height at any given time is dependent on your position which is described in a Cartesian plane by the (x, y) coordinate, thus giving three variables. If you buy a map, you need to know details of the height of the surrounding countryside but you do not want a three-dimensional drawing since this may not be easy to read and would be expensive to produce. The answer is to use contour lines. Lines representing all the points of a fixed height are joined to give contours, so you find your position using the *x*- and *y*-axes and then you look at the nearest contour line to determine your height above sea level. A map of this form is a nomogram, since it shows the relationship between three variables in a single plane. For example, consider the arithmetic mean of two variables, *A* and *B*:

$$\text{mean} = (A + B)/2$$

Since *A* and *B* are variables we can plot them on a Cartesian plane. The equation for the mean forms a straight line linking these two variables; a family of parallel lines can be plotted on the Cartesian plane. This has been done for integer values up to 4 in one quadrant of Figure 6.24.

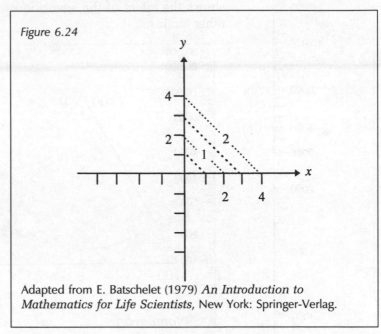

Figure 6.24

Adapted from E. Batschelet (1979) *An Introduction to Mathematics for Life Scientists*, New York: Springer-Verlag.

If you find a point (x, y) (where in this case x and y are both integers), it will lie on a line representing the mean. The value given beside or on this line is therefore the mean of the (x, y) coordinates. More generally, if three variables are related and the relationship is known, two variables can be plotted on the *x,y*-plane as above, with the third being represented by lines on the plane. If you know any two of the three variables, you can therefore find the third. For example, $x = 0$, the mean $= 2$ so y must be 4 since this is the only point all three lines meet on the nomogram in Figure 6.24. You may find that some textbooks call this type of chart a **Cartesian chart** or **concurrency nomogram**.

These charts can often be replaced by **alignment nomograms**, which in their simplest form are simply three vertical lines. Each line has a scale and represents one of the three variables. If you have values for two of the variables, you can join up these points and where it intercepts the third line you have the value of the third variable (Figure 6.25).

One of the most common uses I have found for this is in experimental procedures which involve centrifugation. Centrifugation involves spinning your sample in a circle, at

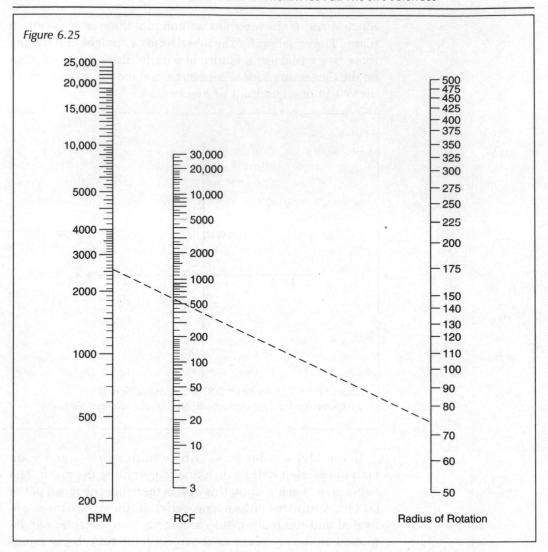

Figure 6.25

speed, so that the centrifugal force causes particles within your sample to pellet. Whether a molecule pellets will depend on its size and the size of the centrifugal force pulling it down. For example, to isolate a cytoplasmic protein from a cell, the cell can be broken open and centrifugation can be used to spin down the heavy cellular debris, leaving the cytoplasmic protein in solution. The protein can then be isolated from the cytoplasmic solution using an appropriate technique.

In scientific papers centrifugation speeds should be given in terms of the number of times the centrifugal force exceeds gravity. This is the relative centrifugal force (RCF) measured in **g**. The size of this force will depend on the radius of the rotor which contains your sample and the spin speed in revolutions per minute. Hence the three variables, force (**g**), rotor radius (mm) and spin speed (rpm) are related; this relationship is given by the equation:

$$\text{rpm} = \sqrt{\frac{\text{relative centrifugal force}}{1.18 \times 10^{-6} \ (\text{radius in mm})}}$$

If you know the radius of the rotor and the centrifugal force required, you can find the spin speed using this equation; but this can be tiresome if it has to be repeated regularly, so an alignment nomogram is often available in the laboratory (Figure 6.25).

A line has been drawn on it to show that a spin of 2500 rpm in a rotor of radius 75 mm would produce a centrifugal force of approximately 600 **g**.

Summary

Data can be presented in the form of tables or figures, but the key rule is to keep the presentation simple so that the trend or observation is clear to the reader. If necessary, the data can be split into two or more groups and represented as a combination of charts and tables. In all cases the data should be given a title to explain what is being shown, with variables and units clearly labelled. The source of the data should be indicated and, where text could clutter the presentation, information can be placed in the figure legend or in footnotes.

Figures can be eye-catching, and visual trends are absorbed faster and more easily from them than by studying the raw data; but in preparing a figure some of the information will be lost, since the actual values for the data points are not quoted. No matter how accurately figures are drawn, data are hard to obtain directly from them with any degree of accuracy. Tables preserve the data but will often need text to point out the trends and guide the reader through the main points.

End of unit questions

1. Protein structures are mainly made up of three secondary structures—the alpha-helix, beta-sheet and turn. Amino acids exhibit preferences for these structures and the conformational parameters give an indication of the preference. The conformational parameters observed for some amino acids in an asparagine-X dipeptide are given in Table 6.19.

   (a) Calculate the relative preference of lysine for the three structures and prepare a triangular chart.

   (b) Use a bar chart to compare the preference of lysine for the three structures.

Table 6.19

X	Helix	Sheet	Turn
Asparagine	0.94	0.18	2.01
Lysine	1.09	0.42	1.84
Arginine	1.97	0.00	0.00

(c) Use a bar chart to compare the probability of finding the amino acids listed in either a beta-sheet or an alpha-helix.

2. The level of protein found within a cell depends on the careful balance between synthesis and degradation. Differences in degradation rates within the cell can therefore have important implications for the regulation of enzyme levels. Those proteins that are rapidly degraded will quickly be removed from the system if synthesis falls and this could be a mechanism of controlling cellular processes. Some of the most rapidly degraded enzymes in the liver have been found to play key roles in metabolic control. The data in Table 6.20 were obtained with respect to degradation of some rat liver proteins. The half-life is the time taken for half the protein to be degraded.

Table 6.20

Protein	Half-life/h
Ornithine decarboxylase	0.2
Cytochrome c	150
Aldolase	118
Tyrosine aminotransferase	2.0
RNA polymerase I	1.2

(a) Sketch the data above as either a histogram or a bar chart.

(b) What made you choose the histogram/bar chart in section (a)?

3. The forced expiratory volume (FEV_1) is used as a diagnostic measure for asthmatic patients. Thirty males were tested and their FEV_1 was represented as a percentage of that expected for a healthy individual. The data are given in Table 6.21.

(a) Prepare a histogram to display the data.

(b) What are the class boundaries you used in (a)?

Table 6.21

FEV$_1$ (% of the expected reading)	Frequency
40–60	3
61–70	3
71–80	8
81–90	9
91–100	5
101–120	2

Source: *Medical Statistics—a Commonsense Approach*, 1993, M. J. Campbell and D. Machin. © John Wiley & Sons (1993). Reprinted by permission of John Wiley & Sons, Ltd.

4. Scientists were interested in the tail length of the snake *Lampropeltis polyzona* as a function of the snake's total length. The length of the tail and the total length were measured (Table 6.22). Decide how you would represent these data in a report and use your method of choice to represent the data. Describe in words the relationship between the two variables given.

Table 6.22

Total length/cm	Tail length/cm
41	5
49	6
81	10
97	12
113	14

Source: Adapted from E. Batschelet (1979), *Introduction to Mathematics*. Berlin: Springer Verlag.

5. The membrane lipid was isolated from the *Escherichia coli* inner membrane. It was found to consist of 75% (w/w) phosphatidylethanolamine, 20% (w/w) phosphatidylglycerol and 5% (w/w) cardiolipin. Use a pie chart to represent the membrane composition.

7 Linear Functions

7.1 Introduction

The aim of many scientific experiments is to try and find the relationship between unknowns. If a variable x is found to be related to a variable y, then it may be possible to describe this relationship mathematically. The rule that would convert x into y is termed a function:

$$x \longrightarrow \boxed{\text{FUNCTION}} \longrightarrow y$$

The simplest form of relationship between two quantities is linear. Linear relationships are found throughout science and have many applications. The objectives of this chapter are:

(a) to introduce the concept of a function;

(b) to introduce proportionality and linear equations;

(c) to show how linear equations can be represented by straight-line graphs.

7.2 Functions

A **function** is a mathematical rule which defines the process by which an input is converted to an output.

$$\text{INPUT} \longrightarrow \boxed{\text{FUNCTION}} \longrightarrow \text{OUTPUT}$$

Let the input be denoted by the letter x. If the we want to double the size of x, the process could be defined by the function as in Example 7.1.

Example 7.1

$$x \longrightarrow \boxed{\text{multiply by two}} \longrightarrow 2x$$

The function would normally be defined mathematically; so if the function was denoted by the letter f, Example 7.1 could be written as:

$$f\colon x \to 2x$$

This reads: 'The function f is such that an input x is converted to the output $2x$'. This is often abbreviated further as shown in Example 7.2, where $f(x)$ indicates that function f acts on the input x.

Example 7.2

$$f(x) = 2x$$

so for $x = 3$

$$f(3) = 2 \times 3 = 6$$

Notice that the input is a variable and the output is also a variable. If we define the output as variable y, then the function $f(x)$ can be written as:

$$y = 2x$$

This equation describes the function in Example 7.2; x is said to be an **independent variable** since its value is freely chosen, and y is the **dependent variable** since the value of y depends on which value of x was used as the input. Remember from Chapter 3 that with algebraic notation the letters are not important so long as they are fully defined; the same applies to functions. $f(x)$ could have written as:

$$h(t) = 2t$$

This would read: 'Function h acts on input t to produce output $2t$'.

Notice that in the above example each value of input produces exactly one output, as shown in Example 7.3.

Example 7.3

$$h(6) = 12$$

It is essential when defining a function that the statement in Box 7.1 is met.

Box 7.1

A function is a mathematical rule which produces a *single* output, y, for each input, x.

To define a function we need to know what the function does, but we also need to know what group of inputs it will act on. The group of all possible inputs is termed the **domain**. Since each input will be mapped onto a single output, this will produce a set of outputs that are linked to the inputs by the function. The set of outputs is called the **range**. To fully define a function you need to state:

(a) A set of numbers which will act as the domain. If a domain is not given it is taken to be the set of all real numbers (Section 1.2).

(b) The set of numbers which form the range.

(c) The rule which allows each member of the domain to be associated with a single member of the range.

Sometimes a function uses different rules for different intervals, as shown in Example 7.4.

Example 7.4

$$f(x) = \begin{cases} x+2 & 0 \le x \le 5 \\ 2x & x > 5 \end{cases}$$

so

$$f(4) = 4 + 2 = 6 \quad but \quad f(10) = 2 \times 10 = 20$$

Worked examples 7.1

(a) $f(x) = 2x^2 - 5$ such that $-5 < x < 5$
Evaluate $f(4), f(0), f(-3), f(5)$.

(b) Write the following in the form of a function:

(i) three times the input minus two and the result is divided by six

(ii) the input is squared, multiplied by five and then subtracted from eight.

7.2.1 Inverse functions

Suppose there is a function $f(x)$ which is defined so that it converts the input x to the output y. A second function is found which takes the output y from $f(x)$ and converts it back to x. This second function is called the inverse of $f(x)$.

$$x \longrightarrow \boxed{\begin{array}{c} \text{Function} \\ f(x) \end{array}} \longrightarrow y \longrightarrow \boxed{\begin{array}{c} \text{Inverse of} \\ f(x) \end{array}} \longrightarrow x$$

If the inverse is given by $h(y)$, it is important to realise that its domain is the range of $f(x)$ since it is accepting as input all the output from $f(x)$. Since the inverse is producing the original x values as its output, the range of $h(y)$ is the same as

The superscript in
$f^{-1}(x)$ shows that this
is an inverse
function—it is not a
power term

the domain of $f(x)$. This is important because the inverse must be able to convert all the outputs from $f(x)$ back to their original form. Although the input variable can be denoted by any letter it is usual to use x. The inverse of $f(x)$ is often defined as $f^{-1}(x)$, where the superscript -1 indicates that this is the inverse of $f(x)$.

Box 7.2

Let $f(x)$ have a domain denoted by A and a range denoted by B; then its inverse function $f^{-1}(x)$ would have domain B and range A.

The statement in Box 7.2 is illustrated by Example 7.5.

Example 7.5

Let $f(x) = 2x$ with $\{0 \le x \le 5\}$
This can be shown on a graph by plotting some values for (x, y) where y is the output, i.e.

$$y = f(x) \quad so \quad y = 2x \ (see \ Table \ 7.1)$$

It can be seen from plotting a few points that the range of $f(x)$ would be $\{0 \le x \le 10\}$.

Table 7.1   Data for x and y such that $y = 2x$

x	0	1	2	3	4	5
$y = 2x$	0	2	4	6	8	10

Figure 7.1   **The function $f(x) = 2x$.**

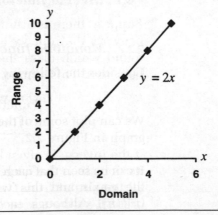

In this case we can define the inverse function $f^{-1}(x)$ where:

$$f^{-1}(x) = \frac{x}{2} \quad \text{with} \quad 0 \le x \le 10$$

It can be shown that $f^{-1}(x)$ is the inverse of $f(x)$ by inserting the output of $f(x)$ into $f^{-1}(x)$:

$$f(1) = 2 \qquad f^{-1}(2) = 1$$

or more generally,

$$f(x) = 2x \qquad f^{-1}(2x) = x.$$

In general, to find an inverse function you need to follow three steps:

(a) express the function $f(x)$ in the form $y = f(x)$;

(b) transpose the formula to make x the subject;

(c) interchange the symbols x and y.

Example 7.6

Find the inverse of $f(x) = 5x + 3$

(a) $y = 5x + 3$ *Convert $f(x)$ to $y = f(x)$*

(b) $x = \dfrac{(y - 3)}{5}$ *Make x the subject*

(c) $y = \dfrac{(x - 3)}{5}$ *Exchange x and y*

so $f^{-1}(x) = \dfrac{(x - 3)}{5}$

Worked examples 7.2

Find the inverse of the following functions:

(i) $f(x) = 7x + 3$ (ii) $f(x) = 3 - x$ (iii) $y = \dfrac{1}{x}$ (iv) $t(x) = \dfrac{-5}{6x}$

7.2.2 *Monotone functions*

Consider the following function:

$$f(x) = x^2 \quad \text{such that} \ -5 \le x \le 5$$

We can plot some of the points (listed in Table 7.2), to get the graph in Figure 7.2.

It can be seen that each value in the range of $f(x)$ is shown on the y-axis and this will form the domain of the inverse $(f^{-1}(x))$. Although each value in the domain of $f(x)$ maps

Table 7.2   Data for x and y such that $y = x^2$

x	-5	-4	-3	-2	-1	0	1	2	3	4	5
$y = x^2$	25	16	9	4	1	0	1	4	9	16	25

Figure 7.2
Plot of the function $f(x) = x^2$.

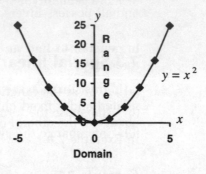

onto only one point in the range, there is a problem if you try to reverse this process since each y value maps onto two values on the x-axis. This is because $f(x)$ maps two values in its domain onto one point in the range, as shown in Example 7.7.

> If more than one point in the domain maps to the same point in the range, then the function does not have an inverse

Example 7.7

$f(x)$ maps each x onto one output:

$$f(-2) = 4$$

but more than one input gives the same output:

$$f(2) = f(-2) = 4$$

The inverse of $f(x)$ would therefore have to map one point in its domain, given by the y-axis in Figure 7.2, onto two points in the range. By definition, a function must map each point in the domain onto only one point in the range so in this case, since this is not possible, $f(x)$ does not have an inverse function.

   If a function maps more than one element in the domain to the same element in the range, an inverse cannot be found. If, for example, a function maps three different x values from the domain onto a point y in the range, how can you determine which of the three x values to map y back to? The easiest way to find out whether there is an inverse function is to plot a graph as in Figure 7.2; but in general if two points in the domain which we will call x_1 and x_2 are such that

$x_1 > x_2$, then if for all cases $f(x_1) > f(x_2)$, the function is said to be **monotonically increasing** and will have an inverse. The graph of $y = 2x$ in Figure 7.1 shows a monotonically increasing function. Similarly, if for all cases of $x_1 > x_2$ we have $f(x_1) < f(x_2)$, this is a **monotonically decreasing** function and will have an inverse. Logarithmic and exponential functions are monotone functions, as are linear functions. These functions are discussed in the following chapters and all possess inverses.

7.3 Special linear equations

In this situation the two quantities under investigation are related and a fixed change in one quantity leads to a fixed change in the other. For example, consider x and y as shown in Example 7.8.

Example 7.8

$$y = 2x$$

In this example, each time there is a set change in y there is a set change in x, but in the case of y the magnitude of the change is two-fold greater than that of x. The equation in Example 7.8 can be rearranged so that x and y can be represented as a **proportion**:

$$y/x = 2$$

Any two quantities that can be represented as a proportion such that:

$$y/x = \text{a constant}$$

are said to be proportional. The symbol of proportionality is '$\propto$' and this indicates that the two sides of the equation are not equal but that they are related and can be represented as a proportion (Box 7.3).

Box 7.3

$A \propto B$
implies $A = kB$ where k is a constant

It can be seen from Box 7.3 that a proportionality can be turned into an equality by the insertion of a constant. The constant is termed the **constant of proportionality**. One of the simplest relationships possible can therefore be described by the equation in Box 7.4, which is said to be a **special linear equation**. If plotted on a graph in a Cartesian plane as shown in Figure 7.3, this type of equation will always produce a straight line that passes through the origin and which has gradient m (Section 6.3.6.1). An example of a special linear relationship is plotted in Figure 7.3; the data are set out in Table 7.3.

Box 7.4

$$y = mx$$

Table 7.3   Data for x and y such that $y = 3x$

x	-1	0	1	2	3
y	-3	0	3	6	9

Figure 7.3
Plot of x and y such that $y = 3x$.

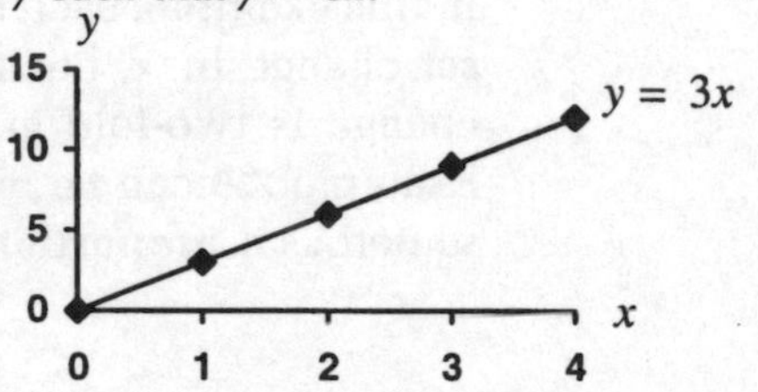

The graph is a straight line passing through the origin so y is proportional to x, i.e.

$$y \propto x$$

or $$\frac{y}{x} = \text{a constant}$$

If you choose any of the ordered pairs from Table 7.3, the constant is found to be equal to 3.0. Suppose you start at a point (x_0, y_0) and move to a point (x, y) where $x > x_0$. The gradient is given by the following equation:

$$\text{gradient} = \frac{y - y_0}{x - x_0}$$

Any two points that lie on the line in Figure 7.3 can be used to find the gradient, as in Example 7.9.

Example 7.9

$$(x_0, y_0) = (2, 6) \quad and \quad (x, y) = (5, 15)$$

$$\text{Gradient} = \frac{(15 - 6)}{(5 - 2)} = \frac{9}{3} = 3$$

Example 7.9 confirms that if you have two variables that are related by a special linear equation, the constant of proportionality is equal to the gradient.

7.4 General linear equations

Look at Figure 7.4.

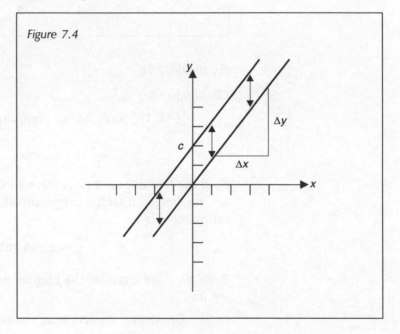

Figure 7.4

The lower line can be represented by a special linear equation as described above. The upper line represents a linear relationship between x and y since it produces a straight line, but the line does not pass through the origin so the relationship is not described by the special linear equation. The upper line crosses the y-axis at $(0, c)$ and so has y-intercept c. The upper and lower lines are parallel, so their slopes are the same. Both lines therefore have gradient m. Consider the example shown in Figure 7.5, which is plotted from the data in Table 7.4.

Taking any two points from Table 7.4, we find that the ratio x/y is no longer constant (Example 7.10).

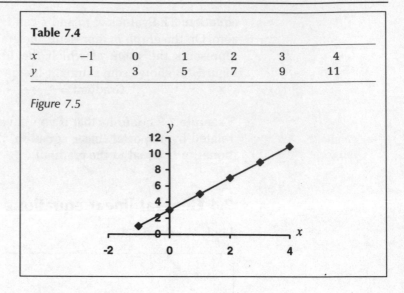

Table 7.4

x	-1	0	1	2	3	4
y	1	3	5	7	9	11

Figure 7.5

Example 7.10

Consider

$$(-1, 1) \quad and \quad (3, 9): comparing\ the\ ratios\ x/y$$

$$\frac{-1}{1} = -1 \quad and \quad \frac{3}{9} = \frac{1}{3}$$

So if the line does not pass through the origin, the variables are no longer directly proportional since they cannot be represented by:

$$\frac{x}{y} = a\ constant$$

Even so, if we consider the change in x and the change in y we find:

$$Gradient = \Delta y / \Delta x = m \quad where\ m\ is\ a\ constant$$

So changes in rate are proportional. Any straight line can be represented by the general linear equation of the form shown in Box 7.5.

Box 7.5 **General equation of a straight line.**

$y = mx + c$ where m and c are constants

This relationship is represented in Figure 7.4, where the upper line is obtained by adding c to all the points in the lower line. The lower line is described by the special linear equation. The values of m and c can be positive, negative or zero. On the graph m represents the gradient of the line and c represents the point at which the line crosses the y-axis. Figure 7.6 shows some straight-line graphs.

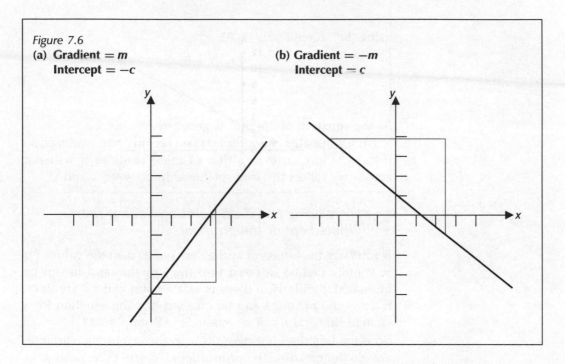

Figure 7.6
(a) **Gradient** $= m$
    Intercept $= -c$

(b) **Gradient** $= -m$
    Intercept $= c$

7.4.1 *Determining the equation of a straight line*

Many computer packages will plot the data, produce a line of best fit by linear regression and supply the equation of the line—but suppose you need to find the equation of the line yourself. This can be calculated as follows.

7.4.1.1 *By using two data points*

Suppose the line passes through the points $(0, 3)$ and $(3, 9)$, then the gradient is given by:

$$\text{Gradient} = \frac{y - y_0}{x - x_0}$$
$$= \frac{(9 - 3)}{(3 - 0)}$$
$$= \frac{6}{3}$$
$$= 2$$

The equation of a straight line is $y = mx + c$ so if we substitute for x, y, and m using the gradient from above and the values from one of the readings:

$$y = mx + c$$
$$y = 2x + c$$

using the ordered pair $(3, 9)$:

$$9 = (2 \times 3) + c$$
$$c = 9 - 6$$
$$= 3$$

So the equation of the line is given by: $y = 2x + 3$

Notice that this equation is based on only two readings, so if there is any error in either of these readings it will not accurately reflect the true relationship between x and y.

7.4.1.2 By using one data point and either the intercept or the gradient

If you have the intercept and a data point, then the values for x, y and c can be inserted into the equation and m can be evaluated. Similarly, if there is a data point and the gradient, the x, y and m values can be inserted into the equation for a straight line and c can be evaluated (Example 7.11).

Example 7.11

You know that the point $(7, 3)$ lies on the line and that $m = 4$, so:

$$y = mx + c$$
$$3 = (4 \times 7) + c$$
$$c = 3 - 28 = -25$$

Notice that again you are reliant on one data point, so any error in the reading will affect the derivation of the equation.

7.4.1.3 From a data set

A line of best fit will average out the error in the data

If you have a range of data readings, any two could be used to obtain the equation of the line linking them as described in Section 7.4.1.1. Ideally the data set should be plotted and a line of best fit placed through the data. This line should generally have the same number of points above it and below it since this acts to 'average out' any error in the readings. The gradient can be measured and the y intercept recorded. This will give the equation of the line.

Worked examples 7.3

Asuming two variables x and y are linked by a linear relationship, find the equation of the line from the following data.

(i) $(0, 2)$ and $(2, 5)$ fall on the line
(ii) The line has a y intercept 3 and includes the ordered pair $(4, 4)$.

7.5 Solving linear equations

You may find the general linear equation written as:

$$0 = ax + b$$

where a and b represent constants and x is a variable. To solve this equation all the possible values for x must be elucidated. In the case of a linear equation, the solution simply requires transposition of the equation to make x the subject (Example 7.12).

Example 7.12

$$2x - 6 = 2 \quad \text{Note that this is the same as } 2x - 8 = 0$$
$$\text{where } a = 2, b = -8$$
$$2x = 8$$
$$x = 4$$

This result can be confirmed by substituting for x *in the original equation:*

$$2 \times 4 - 6 = 2$$
$$8 - 6 = 2 \quad \textit{which is correct.}$$

If there is more than one term containing x, then simply collect all the like terms together before transposing the equation (Example 7.13).

Example 7.13

$$3x - 2x + x = 2 \quad so \quad 2x = 2$$
$$x = 1$$

Worked examples 7.4

Solve the following for the unknown:

(i) $t - 3 = 0$ (ii) $5a = 2a + 3$ (iii) $x + 9 = 2$
(iv) $2 = \dfrac{1}{(x - 7)}$.

7.6 Biological applications

There are many situations in the life sciences that can be described by a linear relationship but there are also many nonlinear relationships which are transformed into a straight-line form to aid analysis. We will consider an example of each in this section.

7.6.1 Beer–Lambert law—an example of a special linear equation

The Beer–Lambert law is a combination of the fundamental principles of absorption spectroscopy. It is based on the principle that certain compounds will absorb light of a given wavelength and that the level of absorption is directly proportional to concentration of material present. Consider vitamin B_2, which absorbs light at 260 nm. Light can be considered to consist of discrete particles called photons. If each molecule of vitamin B_2 can absorb one photon of light, then the more molecules we have present the more light is absorbed:

$$A \propto c$$

where A represents absorption at a set wavelength and c is concentration in molar units. Usually the light is passed through a cuvette of 1 cm length, but if the light were passed through 2 cm of vitamin B_2 solution then obviously it would pass through more material and more light would be absorbed. In fact,

$$A \propto l$$

where l is the path length in cm. This is often set to the value of 1. In fact the Beer–Lambert law shows that

$$A \propto cl$$

Since this is a proportionality, it is known that

$$\frac{A}{cl} = a$$

where a is a constant, and this can be used to determine the concentration of an unknown, given the absorption of a known standard (Example 7.14).

Example 7.14

A 1 cm *path length is used and a* 0.1 mM *solution of dye is found to absorb* 0.6 *units of light at* 595 nm. *A solution of unknown concentration absorbs* 0.3 *units of light at* 595 nm. *What is its concentration?*

$$\frac{A}{cl} = a \quad and \quad \frac{0.6}{0.1 \times 1} = 6\,\text{mM}^{-1}\,\text{cm}^{-1}$$

For the other sample we must have

$$\frac{A}{cl} = 6 \quad so \quad \frac{0.3}{c \times 1} = 6$$

$$c = \frac{0.3}{6} = 0.05\,\text{mM}$$

Although proportionality can be used as in Example 7.14 it is a poor method since it relies on the accuracy of your initial reading. It would be better to take a number of readings for samples of known concentration and to use these data points to prepare a calibration curve, as demonstrated in Example 7.15.

Example 7.15

Trypsin is an enzyme which degrades proteins. BAPA is a chemical which can be used as an artificial substrate for this enzyme since trypsin converts it to para-nitrophenol which is yellow and absorbs light at 410 nm. *To determine how much para-nitrophenol is produced under given conditions a calibration curve is prepared by measuring the absorption of known concentrations of nitrophenol. The data for the curve are presented below.*

Table 7.4 Data for a calibration curve for *para*-nitrophenol

Concentration of *para*-nitrophenol (mM)	Absorbance reading at 410 nm
0	0.00
20	0.06
40	0.13
60	0.20
80	0.27
100	0.33

Source: Data are fictitious.

Notice that in the chart in Figure 7.7 the line ends at the last point so it does not continue past the value given by the last piece of data. Extending the line without data is termed **extrapolation** and should be avoided, since you have no experimental evidence to show that the relationship continues in a straight line past that last point. Calibration

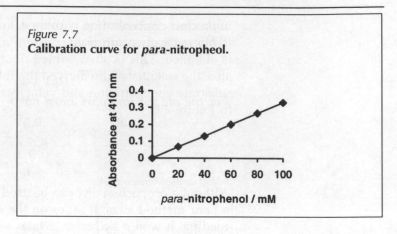

Figure 7.7
Calibration curve for *para*-nitropheol.

curves are widely used in life sciences: known quantities are measured to give a standard curve, and this is used to determine an unknown. For example, using Figure 7.7 an absorption of 0.18 at 410 nm gives the concentration of the unknown as 54 mM.

We have stated that $A/cl = a$ where a is the constant of proportionality, which in this case is termed the molar extinction coefficient and is denoted by ε. This represents the amount of light absorbed by a 1 M solution of the compound at a given wavelength. This allows for the fact that the level of light absorption varies with wavelength and compound. The Beer–Lambert Law is usually given as:

$$A = \varepsilon c l$$

Worked examples 7.5

A solution absorbs 0.6 units of light at a given wavelength. What would the absorption be if:

(i)   The solution was diluted three-fold?

(ii) The path length was increased two-fold?

(iii) The path length was inceased four-fold and the solution was diluted four-fold?

7.6.2   *The Lineweaver–Burk plot*

A simple enzyme-catalysed reaction which converts substrate (S) to product (P) can be modelled by the reaction:

$$S + E \rightleftharpoons ES \rightarrow P$$

where E represents the enzyme and ES the enzyme–substrate complex. The rate of reaction can be measured as the change in product concentration with time, and if the rate of reaction is plotted against substrate concentration a hyperbolic curve is obtained. This is often termed the **Michaelis–Menten plot** after the scientists who derived the relationship between the substrate concentration and velocity. An example is shown in Figure 7.8.

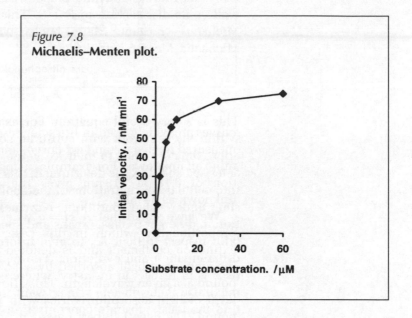

Figure 7.8
Michaelis–Menten plot.

The plot in Figure 7.8 is of a hyperbolic curve and has the equation:

$$y = \frac{ax}{b + x}$$

where a and b are constants. Henri, Michaelis and Menten discovered that by making some assumptions regarding the reaction mechanism, the constant a could be taken to be the maximum velocity that the reaction can obtain under those conditions (V_{max}). If the instantaneous or initial velocities are measured, then there is little change in substrate concentration and constant b is found to represent the dissociation constant for ES, i.e. it gives a measure of the rate at which ES converts back to $E + S$. The Michaelis–Menten equation therefore states that under certain conditions the rate of reaction changes in response to substrate concentration according to the equation:

$$v = \frac{V_{max}[S]}{K_s + [S]}$$

where v is velocity in $(\text{mol litre}^{-1}\,\text{min}^{-1})$, [S] is substrate concentration (M), K_s is the dissociation constant (M) and V_{max} is the maximum velocity obtainable $(\text{mol litre}^{-1}\,\text{min}^{-1})$. Briggs and Haldane used a different approach to derive the Michalis–Menten equation and with different assumptions found a slightly different value for constant b. This derivation took into account the breakdown of ES to product as well as its dissociation to E + S. This is termed K_m or the Michaelis constant. Many books will simply state the Michaelis–Menten equation as

$$v = \frac{V_{max}[S]}{K_m + [S]}$$

This is a very useful equation. For example, most enzymes within the cell work with substrate concentrations that are approximately equal to their K_m value. If an enzyme is characterised and its K_m is determined, this can give some insight into cellular concentrations of metabolites. Since K_m is constant under stated conditions, enzymes with the same function in different tissues, organs and species can be compared with respect to their K_m to give information regarding cell differentiation and evolution. The equation also has many uses with respect to industry, where companies wanting to maximise their efficiency may want to model a reaction to find the ideal substrate concentration for the process being developed.

Unfortunately V_{max} can only be estimated from the above equation, since a hyperbolic curve will tend to V_{max} but will never reach it under experimental conditions. In Figure 7.9(a) the plot appears to reach V_{max}; but if that part of the curve is enlarged, it is seen to be still approaching V_{max} as indicated in Figure 7.9(b).

To find V_{max} it is necessary to transform the hyperbolic plot into a straight-line form. There are a number of ways in which this can be done. One of the most common methods used is that developed by Lineweaver and Burk. This involves rearranging the equation and taking reciprocals:

$$v = \frac{V_{max}[S]}{K_m + S}$$

$$\frac{1}{v} = \frac{K_m + [S]}{V_{max}[S]} \qquad Invert$$

$$= \frac{K_m}{V_{max}[S]} + \frac{[S]}{V_{max}[S]} \qquad Separate\ out\ terms$$

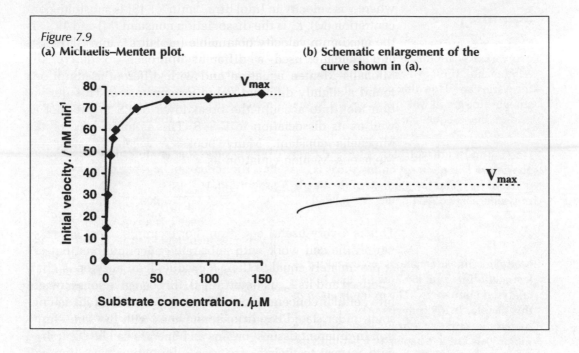

Figure 7.9
(a) Michaelis–Menten plot.

(b) Schematic enlargement of the curve shown in (a).

i.e.

$$\frac{1}{v} = \frac{K_m}{V_{max}} \times \frac{1}{[S]} + \frac{1}{V_{max}}$$

This equation now has the form $y = mx + c$ where

$$y = \frac{1}{v}; \quad x = \frac{1}{[S]}; \quad m = \frac{K_m}{V_{max}}; \quad c = \frac{1}{V_{max}}$$

The graph of $1/v$ against $1/[S]$ will therefore produce a straight line and the y intercept is equal to $1/V_{max}$ so V_{max} can be found. The gradient is K_m/V_{max} so, if V_{max} is known, K_m can be found.

Worked examples 7.6

Multiplying both sides of the Lineweaver–Burk equation by [S] gives rise to the Hanes–Woolf equation. This will also provide a straight-line graph.

(a)   What is the Hanes–Woolf equation?

(b)   What would you plot?

(c)   How would you determine K_m and V_{max}?

Summary

There are many cases in life sciences where two variables are linked by a linear relationship. If plotted as a function $f(x)$ on the Cartesian plane, this will produce a straight-line graph which is described by the general linear equation:

$$y = mx + c$$

where y and x are variables and m and c are constants. The intercept on the y-axis is denoted by c and the gradient of the line is m where for two data points (x_0, y_0) and (x, y), $x > x_0$,

$$m = \frac{y - y_0}{x - x_0}$$

Many non-linear equations are converted to a straight-line form to allow the data to be analysed more easily. The example given in this chapter is the conversion of a hyperbolic curve to a straight line by taking reciprocals but in Chapter 10 logarithms are shown to be regularly used to convert power functions and exponential functions to straight-line forms.

If the data produce a straight line which passes through the origin, then the y intercept is zero and the equation reduces to that of the special linear equation:

$$y = mx$$

Any variables that are linked by a special linear function can be represented as a proportion, so

$$A \propto B \text{ implies } A = kB$$

where k is a constant.

If you have an ordered pair (x, y) with $x \propto y$, the constant k can be determined and this can be used to form ordered pairs from values of x alone or y alone, although this method relies on the accuracy of the initial reading. Special linear equations are very common in the preparation of calibration curves.

End of unit questions

1. Five strains of bacteria are numbered one to five and are exposed to a potential antibacterial. If the strain is killed, the result is recorded as one. If the bacteria survive, the result is recorded as zero. Could this be considered to be a function?

2. The amount of light absorbed by a dye is directly proportional to the concentration of dye present. The process can be modelled using a special linear equation and a 5 mM solution absorbs 0.45 absorption units at 540 nm.

    (a) Assuming the data point is accurate, what would you expect the equation of the calibration curve to be?

    (b) Use the idea of proportionality to determine the concentration of a dye solution that absorbs 0.2 units at 540 nm.

    (c) Could you use the data given to determine the concentration of a solution that absorbs 0.78 units at 500 nm?

3. A 2% (w/v) solution of riboflavin (vitamin B_2) absorbs 0.48 units of light at 444 nm and a path length of 1 cm. Use this information to answer the following:

   (a) What is the absorption of a 6% (w/v) solution at path length 0.5 cm?

   (b) What is the concentration of a solution with absorption 0.12 units at 450 nm; path length 2 cm?

   (c) The molar extinction coefficient for riboflavin is $12200 \, M^{-1} \, cm^{-1}$. What is the molar concentration of the 2% (w/v) solution?

   (d) Riboflavin has three peaks on its absorption spectra (i.e. a plot of absorption with respect to the wavelength of light). These are characterised in Table 7.5. Assuming you could choose to measure the concentration at any wavelength, which would you choose?

Table 7.5

Wavelength/nm	Molar absorption coefficient/$M^{-1} \, cm^{-1}$
444	12 200
371	10 600
266	27 700

4. The following three data points have been obtained. Without plotting them, confirm that they fall on a straight line. What is the equation of the line?
   $(1.0, 2.0)$, $(3.0, 3.0)$, $(5.0, 4.0)$.

5. The data in Table 7.6 were obtained for an enzyme-catalysed reaction.

   (a) Estimate K_m and V_{max} by deriving the general linear equation that fits the Lineweaver–Burk equation for these data.

   (b) Perform a Lineweaver–Burk plot to find K_m and V_{max}.

   (c) Why is the graphical method considered to be more accurate?

Table 7.6

Substrate concentration/mM	Velocity/nmol litre^{-1} min^{-1}
0.008	13.8
0.010	17.0
0.017	22.3
0.025	30.8
0.050	42.0

8 Power Functions

8.1 Introduction

Power functions were introduced in Chapter 4 where multipliers were used in scientific notation, but powers have many uses and are often found within the life sciences. They are especially relevant in the study of dynamics and rates; for example, the study of blood flow and its effect on artificial hearts would require an in-depth understanding of powers. In this chapter power functions will be discussed and the algebra associated with them will be introduced. Power terms can be combined to give more complex equations called polynomials. Polynomials will be introduced and their relevance to biology demonstrated. The main objectives of this chapter are:

(a)   to introduce power functions and polynomials;

(b)   to provide experience of algebraic manipulation of polynomials.

8.2 Power functions

Power functions are defined by the equation in Box 8.1.

Box 8.1   **Definition of a power function.**

$$y = ax^n$$

where a and n are constants and x and y are variables.

The equation can be seen to imply that y is proportional to x^n (Chapter 7). The behaviour of the function is characterised by the constants a and n and is therefore termed a two-parametric equation. Although it is characterised by two parameters, the behaviour of a power function is mainly governed by the size of the exponent, n. A common use of power functions is the calculation of area and volume. The area of a circle, for example, is given by the equation:

The behaviour of a power term is mainly governed by the size of the exponent

$$A = \pi r^2$$

where A is area (m^2), r is the radius (m) and π represents a constant which can be approximated by 22/7. In this power function $n = 2$, so this is called a **second-degree** function. The volume of a sphere is given by:

$$V = (4/3)\pi r^3$$

where V represents the volume (m^3). The equation for volume is **a power function of the third degree**.

Because second-degree functions are very common, they have their own name and are termed **quadratic functions**. The graph of a simple quadratic function is shown in Figure 8.1.

Table 8.1   Data for the function $f(x) = 2x^2$

x	−3	−2	−1	0	1	2	3
$f(x)$	18	8	2	0	2	8	18

Figure 8.1
Plot of the function $f(x) = 2x^2$.

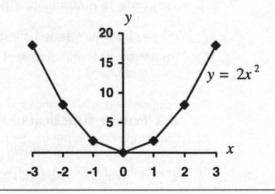

A graph of the special quadratic function (Box 8.2) is termed a **quadratic parabola** and Figure 8.1 is an example of a parabola.

Box 8.2   **The special quadratic function.**

$$y = ax^2$$

where a is a constant and y and x are variables.

The parabola is symmetrical about the y-axis and it passes through the origin.

8.3 Polynomials

A quadratic function such as

$$y = ax^2$$

can be written in the form of a **general quadratic equation**. This is shown in Box 8.3.

Box 8.3 **The general quadratic function.**

$$y = ax^2 + bx + c \quad a, b, \text{ and } c \text{ are constants; } a \neq 0$$

The right-hand side of the equation in Box 8.3 is said to be a **second-order polynomial in** x. The degree or order of the polynomial is based on the highest power function and, whilst quadratics are probably the most common form, polynomials of the nth degree can also be found in biology. The general quadratic function will produce a parabola in the same way as the special quadratic function $(y = ax^2)$ (Figure 8.1), but the addition of the linear term $(bx + c)$ to the power term means that the vertex no longer has to pass through the origin.

Consider the parabola in Figure 8.1. This represents the equation:

$$y = 2x^2$$

Suppose we wish to move the plot by two units along the x-axis (to the right) and one unit along the y-axis (down), i.e. all the points (x, y) lying on the curve will move to $(x + 2, y - 1)$, so the vertex will move from $(0, 0)$ to $(2, -1)$. If we denote the new points by (X, Y), then we have:

$$X = x + 2$$
$$Y = y - 1$$

To find the equation of this new parabola you need to perform the following operations.

1. Make x and y the subject of the two equations derived for X and Y:

$$X = x + 2 \quad \text{so} \quad x = X - 2$$
$$Y = y - 1 \quad \text{so} \quad y = Y + 1$$

2. Insert these values for x and y in the original equation:
    $y = 2x^2$ becomes $Y + 1 = 2(X - 2)^2$

    so

    $$Y = 2(X - 2)^2 - 1$$
    $$= 2(X - 2)(X - 2) - 1$$
    $$= 2X^2 - 8X + 8 - 1$$
    $$= 2X^2 - 8X + 7$$

    or

    $$y = 2x^2 - 8x + 7$$

So the quadratic term has not changed, but the linear term $-8x + 7$ has been added. This new function is plotted in Figure 8.2.

Table 8.2   Data obtained from $f(x) = 2x^2 - 8x + 7$

x	−1	0	1	2	3	4	5
$f(x)$	17	7	1	−1	1	1	17

Figure 8.2   **Plot of $f(x) = 2x^2 - 8x + 7$.**

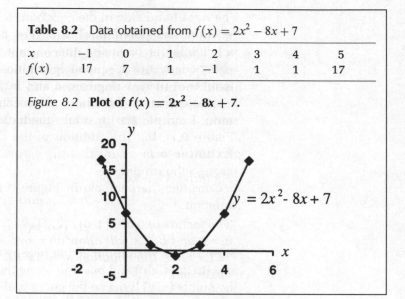

8.4 Solving quadratic equations

A quadratic function will contain the term:

$$y = ax^2$$

It is possible to use a logarithmic transformation to convert power functions into a straight-line form which has a gradient n and a y intercept of $\log a$. This is discussed in Chapter 10, but in this chapter we shall look at algebraic means of solving quadratics.

8.4.1   *Solving by factorisation*

A quadratic expression of the form:

$$ax^2 + bx + c \qquad a \neq 0$$

can be solved algebraically if the equation can be expressed as

$$ax^2 + bx + c = 0$$

and if the left-hand side of this equation can be factorised. There are two possible situations which can occur; these are shown below.

8.4.1.1 *If the coefficient of x^2 is 1*

Consider the expression:

$$(x + a)(x + b) = x(x + b) + a(x + b)$$
$$= x^2 + xb + ax + ab$$
$$= x^2 + (a + b)x + ab$$

It can be seen from the above example that the coefficient of x is equal to $(a + b)$ and the constant is equal to the product ab. To factorise a general quadratic equation we therefore need to find two numbers a and b that will add up to the coefficient of x and whose product produces the constant term. Example 8.1 illustrates the technique.

Example 8.1

Factorise $x^2 + 5x + 6$:

$$a + b = 5 \quad and \quad ab = 6$$

The factors of 6 are $\{1, 6\}, \{2, 3\}, \{-1, -6\}, \{-2, -3\}$, so only $a = 2$ and $b = 3$ will allow $ab = 6$, $a + b = 5$.

To solve the equation $x^2 + 5x + 6 = 0$ the first step is to factorise it:

$$x^2 + 5x + 6 = 0 \quad so \quad (x + 2)(x + 3) = 0$$

It therefore follows that for this to be true either $(x + 2) = 0$ or $(x + 3) = 0$.

This quadratic therefore has two solutions, which are $x = -2$ or $x = -3$.

8.4.1.2 *If the coefficient of x^2 is not 1*

In this case the solution is a little more difficult to find and requires some trial and error. The first step is to try to obtain the quadratic term in its simplest form with the coefficient of x^2 equal to one. Examples 8.2 and 8.3 present two cases.

Example 8.2

$$2x^2 + 10x + 12 = 0$$

so
$$2(x^2 + 5x + 6) = 0$$

The factor in brackets must equal zero for the equation to be true. This is the same as Example 8.1, so $x = -2$ or -3.

It is not always possible to modify the equation such that the coefficient of x^2 is one and this is illustrated in the following example.

Example 8.3

$$4x^2 + 14x + 6 = 0$$

so
$$2(2x^2 + 7x + 3) = 0$$

Inside the brackets, the coefficient of x^2 is two, so the factorised form must be:

$$(2x + a)(x + b) = 0$$

This multiplies out to

$$2x^2 + (a + 2b)x + ab = 0$$

In this case we need to look at all the factors of the constant term, $c = 3$. These factors then need to be investigated by trial and error to find which are the correct factors ab for the expression shown above and which also satisfy the condition $a + 2b = 7$.

Factors of
$$3 = \{1, 3\}, \{-1, -3\}$$

Substituting these factors for a and b in the previous expression $a + 2b$ which is the coefficient of x, the only factors that fit the equation are $\{1, 3\}$. The quadratic can be factorised as

$$(2x + 1)(x + 3) = 0$$

and the two solutions are $x = -\frac{1}{2}$ or $x = -3$

Worked examples 8.1

Solve the following:

(i) $x^2 + 5x - 6 = 0$

(ii) $-2x^2 - x + 3 = 0$

(iii) $x(1 - x) = x(2x - 1)$.

8.4.2 *Solving by using a formula*

Some equations will not factorise and in that case the method described in this section may be more appropriate. In addition, it is often easier to use the formula method for solving quadratics. A quadratic equation of the form:

$$ax^2 + bx + c = 0 \qquad a \neq 0$$

can be solved for x by inserting the values of the constants a, b and c into the following equation:

$$x = -\frac{b \pm \sqrt{b^2 - 4ac}}{2a}$$

Notice that the equation contains the $\pm$ symbol and this gives rise to two possible solutions. One is obtained by using the positive square root and the other by using the negative square root. The square root term in the equation is called the **discriminant** and is denoted by the letter D so:

$$D = b^2 - 4ac$$

The discriminant distinguishes between three possible outcomes:

(a) $D < 0$: the square root is not a real number and so there are no real solutions to the quadratic expression.

(b) $D > 0$: then there are two solutions to the equation x_1 and x_2 and these are termed the **distinct real roots of the equation**.

The discriminant shows whether the equation has zero, one or two real roots

(c) $D = 0$: then both the solutions fall together, i.e. the parabola just touches the x-axis with its vertex so $x_1 = x_2$ and there is only one root, which is termed the **repeated root** or **equal root**.

Worked examples 8.2

Solve the following equations:

(i) $2x^2 - 6x + 4 = 0$

(ii) $x^2 + 4x - 8 = 0$

(iii) $2x^2 - 7x + 3 = 0$.

8.5 Applications in life sciences

Power functions are often found within the life sciences. They are usually involved in equations which describe rates and are most commonly found as quadratic expressions. One example of a higher polynomial is the Hill equation, which is used to estimate the number of binding sites for an allosteric enzyme. The Hill equation is discussed in Chapter 10 so will not be covered here.

8.5.1 Quadratics as a tool to calculate pH

A common use of quadratics is in the study of weak acids. In Chapter 10 we will discuss acidity and pH. pH is simply a measure of the hydrogen ion concentration within the solution. An acid is a compound which releases protons; for a strong acid it is easy to calculate the number of hydrogen ions released, since the acid fully dissociates (Example 8.4).

Example 8.4

Hydrochloric acid is a strong acid and fully dissociates, so 0.1 M *acid produces* 0.1 M *protons:*

$$HCl \rightarrow H^+ + Cl^-$$

With a strong acid, you know the concentration of protons released and so can calculate the pH. Now consider a weak acid. Weak acids are so called because they do not fully dissociate, so 0.1 M acid does not produce 0.1 M protons. Instead, an equilibrium is set up where the rate at which the acid dissociates is equal to the rate at which it is formed so there is no net change in the acid concentration.

Example 8.5

Formic acid is a weak acid and does not fully dissociate so:

$$HCOOH \rightleftharpoons HCOO^- + H^+$$

To calculate the pH the acid concentration (denoted by square brackets, []) is needed but this is unknown so:

Let $\qquad\qquad [H^+] = [HCOO^-] = x\, M$

If the original acid concentration was 0.1 M, *then at equilibrium the acid concentration is given by:*

$$[HCOOH] = (0.1 - x)\, M$$

The amount of product and the amount of substrate are linked by the equilibrium constant, which in this case is the acid dissociation constant K_a, where

$$K_a = ([H^+][HCOO^-])/[HCOOH]$$

Given that the K_a for formic acid is 1.7×10^{-4} M, *we can substitute the values in the equation:*

$$1.78 \times 10^{-4} = (x \times x)/(0.1 - x)$$
$$x^2 = 1.78 \times 10^{-4}(0.1 - x)$$

or $\qquad x^2 + (1.78 \times 10^{-4})x - (1.78 \times 10^{-5}) = 0$

This is now writen in the form of a quadratic and can be solved using the equation given in Section 8.4.2, so $x = 4.2$ mM.

It is worth noting that in this case, because the acid is weak, x is very small compared with the amount of acid present so the above calculation can be simplified by saying

let
$$0.1 - x = 0.1$$

$$1.78 \times 10^{-4} = (x \times x)/(0.1 - x)$$
$$= x^2/0.1$$

so
$$x = (1.78 \times 10^{-4} \times 0.1)^{1/2}$$
$$= 4.2 \text{ mM}$$

It can be seen that in this case the simplification has not affected the answer.

8.5.2 Quadratic equations and rates

There are numerous examples of quadratic equations being used to model rates, such as, blood flow through blood vessels or the diffusion rates of chemotherapeutic drugs into tumours. As an example it has been shown that the rate of photosynthesis in grass is related to temperature and the nature of the relationship is described by the equation

$$y = -0.25x^2 + 6.8x + 46.4$$

where the independent variable x is temperature (°C) and y is the rate of photosynthesis. This equation can be shown to reach a maximum value at $x = 13.6$ and at this temperature the rate of photosynthesis is predicted to be 92% of its maximum by this model.

Summary

Power functions have the form:
$$y = ax^n$$
where a and n are constants. Power functions are described as being of the nth degree; for example, if $n = 2$ the expression would be a power function of the second degree. When $n = 2$ the expression can be called a quadratic function and gives rise to a parabolic plot. If the expression contains more than one power term it is called a polynomial and the behaviour of this form of expression is dependent on the size of the index n. Figure 8.3 shows schematically the form of the plot for second- and third-degree functions and how the index can alter the behaviour of the relationship. Power functions, especially quadratic functions, are widely used in biology and have special relevance in the study of rates.

Figure 8.3

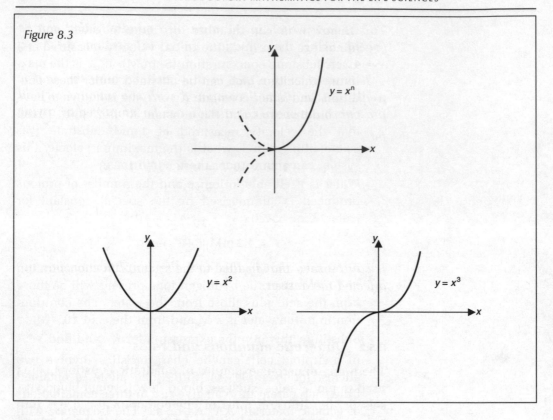

End of unit questions

1. Where possible solve the following equations for x:
    (a) $3x^2 + 4x + 2 = 0$
    (b) $2x^2 - 5x + 3 = 0$
    (c) $2x^2 + 7x + 3 = 0$

2. *Staphylococcus* is a spherical bacterium. Assume it can be modelled by a sphere and that during its life cycle its radius (r) increases by 5%.

    (a) By what percentage must its volume increase? (Volume $= \frac{4}{3}\pi r^3$)

    (b) Given that the surface area is S (cm^2), where $S = 4\pi r^2$, what is the percentage increase in the surface area?

    (b) If the bacterium continues to grow, can you envisage a biological problem?

3. The equation describing the kinetics of a multi-site enzyme-based reaction is termed the Hill equation:

    $$v = \frac{V_{\max}[S]^n}{K' + [S]^n}$$

where n is the number of binding sites and, v (nmol litre^{-1} min^{-1}) is the initial velocity measured at a given substrate concentration [S] (mM). V_{max} is the maximum velocity which can be obtained under these conditions and K' is a constant. An enzyme is known to have two binding sites and the constant K' is $25\,\mu M^2$. Given that the velocity measured at 1 mM substrate was $5\,\mu$mol litre^{-1} min^{-1}, what is the maximum velocity this system can attain under these conditions?

4. Water is itself able to ionise and the number of protons produced is summarised by the special constant for water, K_w:

$$K_w = [H^+][OH^-] = 10^{-14}$$

If a strong acid is added to the system at a concentration of 10^{-6} M, then the hydrogen ions present will be those from the acid plus those from the water. The contribution from the water is x M and from the acid 10^{-6} M. Substitute this into the expression for K_w and find x.

5. In a diploid cell, genetic characteristics involve two alleles, for example Aa. Since one allele is obtained from each parent this can give rise to three combinations of genetic material: AA, Aa or aa. The frequency with which these occur is described by the Hardy–Weinberg equilibrium, which says that in a population the frequency of AA, is p^2, that of Aa is $2pq$, and q^2 for aa. p and q are obtained as follows:

$$\text{Allelic frequency of A} = p = \frac{(2AA + Aa)}{(2(AA + Aa + aa))}$$

$$\text{Allelic frequency of a} = q = \frac{(2aa + Aa)}{(2(AA + Aa + aa))}$$

In a population, 58 individuals with the AA genotype are noted, 37 with Aa and 5 with aa. Is this population at the Hardy–Weinberg equilibrium?

9 Exponential Functions

9.1 Introduction

In the life sciences many measurements that are recorded relate to the growth of an organism or population. It is obvious that living organisms do not change in regular steps, i.e. the change is not modelled by a linear process. The rate at which a population will expand is related to the size of the population. The more parents there are, the more children will be born, so as the population gets bigger the growth rate increases. Growth patterns like these can be described by using exponential functions and these will be introduced in this chapter. First, the idea of numerical sequences will be explored, differentiating between arithmetic sequences, which change in regular steps, and geometric sequences, where the size of the change depends on the size of the value that is changing. The aims of the chapter are:

(a)   to introduce arithmetic and geometric sequences;

(b)   to introduce exponential functions;

(c)   to show the importance of geometric sequences and exponentials in the life sciences.

9.2 Sequences

A numerical **sequence** refers to an ordered arrangement of values. Suppose we select three consecutive numerical values $\{1, 2, 3\}$ and apply a function $f(x)$ to each. If the output of the function is arranged:

$$f(1), f(2), f(3)$$

this forms a sequence. Notice that to form the sequence the variable y, where $y = f(x)$, must be arranged in order after the input of *consecutive* integer values.

9.2.1 *Geometric sequences*

Consider the division of a bacterial cell. Each cell divides to give two daughter cells and for the Gram-negative bacterium *Escherichia coli* the time taken for the population to double in this way is approximately 20 min. Assume that at time zero we have one cell, therefore after one doubling time (20 min) we have two cells and after a further doubling time each of these has divided to give four cells. This could be modelled by the following function:

$$N = 2^t \qquad t \in \{0, 1, 2, 3, 4 \dots\}$$

where t is the number of doubling times that have elapsed and N is the number of cells. Using the function $f(t) = 2^t$ and arranging the outputs in order gives the following sequence:

$$1, 2, 4, 8, 16, 32 \dots$$

In this example, the independent variable is an **exponent**; there are many relationships within life sciences where this occurs. The sequence generated from an exponential function is said to be a **geometric sequence**. A geometric sequence is therefore created by the output of a function $f(x)$ such that:

$$f(x) = ta^x \qquad a > 0 \quad x \in \mathbb{R}$$

The general form of a geometric sequence would therefore be:

$$t, ta, ta^2, ta^3, ta^4 \dots$$

If you take any two consecutive terms in a geometric sequence you will find they always give the same ratio (Examples 9.1 and 9.2).

Example 9.1

Taking ta^n and $ta^{(n+1)}$
Then the ratio is given by $\quad ta^{(n+1)}/ta^n = a^{(n+1)}/a^n$
$$= a$$

Any two consecutive values in a geometric sequence will give a common ratio

The value of a is therefore said to be the **common ratio**.

Example 9.2

Suppose you have a sequence where consecutive values are linked by a common ratio, such as:

$$100, 140, 196$$

This sequence was simply obtained by multiplying each of the preceding values by 1.4. *The ratio of any two consecutive values is therefore* 1.4:

$$140/100 = 1.4$$

The arithmetic mean (Chapter 11) of two numbers is given by:

$$\text{Mean} = (x_1 + x_2) \div 2$$

This is the value that lies on the number line halfway between x_1 and x_2. For example, using the sequence in Example 9.2, the arithmetic mean of 100 and 196 is given by $(100 + 196) \div 2 = 148$. This gives us the value that is equidistant from both numbers. Yet in this case the sequence is geometric, not linear, so it would be better if the mean was nearer to 140 since this is the middle value in the sequence. If a sequence contains numbers linked by a common ratio you can use the **geometric mean** to find the value which would occur between x_1 and x_2, rather than the arithmetic mean given above. The geometric mean x_g is given by:

> The geometric mean of two values finds the number which would lie between them in a geometric sequence

$$x_g = \sqrt{(x_1 x_2)}$$

This equation is used in Example 9.3.

Example 9.3

Using data from the sequence in Example 9.2,

$$x_g = (100 \times 168)^{1/2} = 140$$

Within any geometric sequence the geometric mean of the values at positions $n - 1$ and $n + 1$ is always equal to the value occurring at position n, i.e.

$$x_g = x_n = \sqrt{(x_{n-1} x_{n+1})}$$

In Example 9.3 the geometric mean of the numbers at positions one and three gave the value of the number at position two.

9.2.2 *Arithmetic mean*

We briefly introduced the concept of an arithmetic mean in Section 9.2.1 and this is discussed more fully in Chapter 11. In the same way as a geometric series involves exponential functions, arithmetic sequences involve linear functions but in the latter case it is the difference between two consecutive terms that is constant. An example of an arithmetic sequence is given by the function:

$$f(x) = ax + b \qquad x \in \{0, 1, 2, 3, 4, \ldots\}$$

so the sequence is:

$$b, a + b, 2a + b, 3a + b, 4a + b$$

In this case we do not have a constant ratio, but we do have a constant difference since in all cases:

$$(ax_{n+1} + b) - (ax_n + b) = ax_{n+1} + b - ax_n - b$$
$$= a$$

where a is termed the **constant difference**. To find what value occurs in the sequence between values $(n-1)$ and $(n+1)$ you would take the arithmetic mean, since the arithmetic mean of the values at positions $(n-1)$ and $(n+1)$ always gives the value of the factor at position n in an arithmetic sequence.

9.3 Exponential functions

For exponential functions the exponent is the independent variable but for power functions the exponent is a constant

An exponential expression contains a variable x in the form a^x where a is a constant. The constant a is termed the base and x is the power or index. Be very careful not to confuse exponential functions with power functions. With the power function in Chapter 8 the variable was the base and the exponent was formed by a constant (Box 9.1).

Box 9.1

> Power function $\quad x^a$
> Exponential function a^x

On your calculator you will find e^x. This is one of the most commonly used exponential expressions in life sciences since it describes many natural phenomena, such as population growth and radioactive decay. The value of e to four decimal places is 2.7183. Exponential expressions obey the normal rules of algebra discussed in Chapter 3 and in addition they obey the laws of indices described for powers in Chapter 4, as demonstrated in Example 9.4.

Example 9.4

$$e^3 \times e^2 = e^5 \qquad \textit{Using power rules}$$

Worked examples 9.1.

(a) Simplify the following equations:
   (i) $e^{4x}(e^{x/2} + e^{-x})$ (ii) $\sqrt{e^{8x}}$ (iii) $(e^x + e^{7x})/e^{2x}$
   (iv) $e^x - (e^{3x})^2$

(b) Calculate the following:
   (i) $e^{1.2}$ (ii) $e^{-0.7}$ (iii) $e^{0.2}$ (iv) $1/e^3$

An exponential function is defined by the general equation:

$$y = ta^x$$

where x and y are variables and a and t are constants such that $a > 0$ and $t > 0$. If $t = 1$, then this simplifies to the special case where:

$$y = a^x$$

Exponential functions possess a number of interesting properties.

(a) The function never becomes negative. As the value of x decreases below zero the output from the exponential function gets smaller and smaller. In other words, as x becomes large and negative the function gradually approaches zero. The function never actually reaches zero and $y = 0$ is termed an **asymptote**.

(b) If x is greater than zero, then as x increases e^x increases rapidly. This is termed **exponential growth**.

For e^x, as the index becomes large and positive the exponential function rapidly increases, showing exponential growth

(c) Since x is an index, if $x = 0$ then $e^x = 1$. A plot of this function therefore always crosses the axis at $y = 1$. A plot for a limited range of x (Table 9.1) is shown in Figure 9.1.

Table 9.1

x	−3.0	−2.0	−1.0	0.00	1.0	2.0	2.5	3.0
e^x	0.05	0.14	0.37	1.00	2.72	7.39	12.18	20.09

Figure 9.1
Plot showing exponential growth $(y = e^x)$.

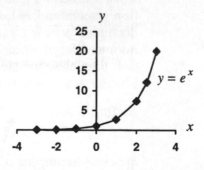

Notice from Table 9.1 how rapidly the function e^x increases with increasing x. This is illustrated more clearly on the plot in Figure 9.1.

This curve can be reflected in the line $y = 0$ by considering $-e^x$ as shown in Figure 9.2.

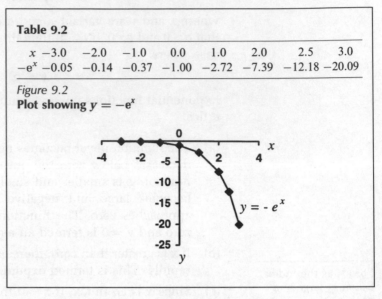

Table 9.2

x	−3.0	−2.0	−1.0	0.0	1.0	2.0	2.5	3.0
$-e^x$	−0.05	−0.14	−0.37	−1.00	−2.72	−7.39	−12.18	−20.09

Figure 9.2
Plot showing $y = -e^x$

A function which is closely related to e^x is:

$$y = e^{-x}$$

As in the case of e^x, the function e^{-x} takes the value one when x equals zero and the function never produces a negative output. In this case, as x gets bigger e^{-x} rapidly decreases (Figure 9.3); this is termed **exponential decay**.

For e^{-x}, as the index becomes large and positive the exponential function rapidly decreases, showing exponential decay

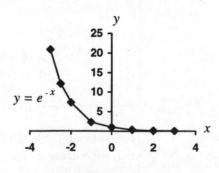

Table 9.3

x	−3	−2	−1	0	1	2	3
$-x$	3	2	1	0	−1	−2	−3
e^{-x}	20.09	7.39	2.72	1.00	0.37	0.14	0.05

Figure 9.3
Plot showing exponential decay ($y = e^{-x}$).

Notice that the data in Table 9.3. follow the same trend as in the case of e^x but that the trend is reversed.

9.4 Solving exponential equations

Exponential equations can be solved by transforming the equation into a linear form using logarithms. This is covered in Chapter 10. Here we will look at a graphical method for solving equations involving exponential terms. For example, consider the equation:

$$e^{2x} - e^{-x/2} + 2 = 0$$

We can rearrange this to place an exponential term on each side of the equation:

$$e^{2x} + 2 = e^{-x/2}$$

If $y = e^{2x} + 2$ and $y = e^{-x/2}$ are plotted on the same x–y plane, then where the lines cross the two expressions are equal so this point is the solution to the equation. If the lines cross at more then one point there is more than one solution. A range of values have been calculated for the two expressions and from the graph in Figure 9.4 the solution is approximately $x = -1.4$.

Worked examples 9.2

Solve the following equations graphically:
(i) $e^x - 2x = 3$ (ii) $e^x - 6.7 = 0$ (iii) $e^x - e^{-x} = 10$

Table 9.4

x	-3	-2	-1	0	1
$e^{2x} + 2$	2.002	2.018	2.135	3.000	9.389
$e^{-x/2}$	4.482	2.718	1.649	1.000	0.607

Figure 9.4

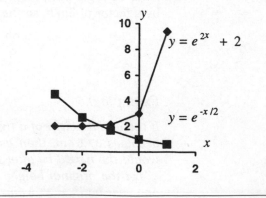

9.5 Applications in biology

Many natural processes that increase or decrease do so according to a geometric sequence, i.e. they follow an exponential process. The use of exponentials is especially common when modelling population growth or decline, since all organisms will tend to increase or decrease at a rate which is proportional to their numbers. Put simply: as the numbers increase, the rate of growth (or decay) increases so can be modelled using an exponential function.

9.5.1 *Exponential growth.*

Suppose you are interested in the growth of a living organism. Growth can often be modelled as an exponential process: If the height of the organism is h_0 cm and after a given time it grows to $h_1$cm, then the new height is:

$$h_1 = (h_0 + \delta h)$$

where δh represents a small change.

If this is an exponential process, the ratio of consecutive terms is constant, i.e.

$$\frac{h_1}{h_0} = c \qquad \text{where } c \text{ is a constant.}$$

so $\qquad h_1 = h_0 c$

Each successive value can therefore be obtained by multiplying the previous value by a constant:

$$h_0 \times c = h_1 = (h_0 + \delta h)$$

$$c = \frac{(h_0 + \delta h)}{h_0}$$

so

$$= 1 + \frac{\delta h}{h_0}$$

After each growth period the organism has increased in size by a factor of $\delta h/h$, so the percentage change would be:

$$\frac{\delta h}{h} \times 100\% = y\%$$

Example 9.5

If the initial height of a tiger is 35.0 cm and after one year it becomes 37.8 cm, then assuming exponential growth what would the height be after five years?

Let the original height $h_0 = 35.0$ cm and the height after one year be $h_1 = 37.8$ cm. If exponential growth is assumed the percentage increase each year will be constant and is given by:

$$\frac{\delta h}{h} \times 100\% = \frac{37.8 - 35}{35} \times 100\%$$

$$= 0.08 \times 100\%$$

$$= 8\%$$

If growth had been given as a percentage, then the question would have stated that the original height was 35 cm and in one year increased by 8%. As a fraction 8% = 8/100 = 0.08. The new height can be represented as:

$$h_1 = h_0 + (h_0 \times y)$$

where y = 0.08, i.e. the original height plus fraction y of that height. This can be factorised (Chapter 3) to give:

$$h_1 = h_0(1 + y)$$

i.e. $$h_1 = 35(1 + 0.08)$$

$$= 35 \times 1.8 = 37.8 \, cm$$

If growth is geometric, then there must be a constant ratio between each successive value in the sequence, i.e.

$$(h_0(1 + y))/h_0 = (1 + y)$$

so each year the increase is obtained by multiplying the preceding value by this ratio:

$$h_n = h_{n-1}(1 + y)$$

i.e. every year the height increases by a factor of y, so we have the sequence:

$$\{h_0\}, \{h_0(1 + y)\}, \{h_0(1 + y)(1 + y)\},$$
$$\{h_0(1 + y)(1 + y)(1 + y), \ldots$$

or the geometric sequence:

$$h_0(1 + y)^0, h_0(1 + y)^1, h_0(1 + y)^2, h_0(1 + y)^3, \ldots, h_0(1 + y)^n$$
$$= h_0, h_0(1 + y)^1, h_0(1 + y)^2, h_0(1 + y)^3, \ldots, h_0(1 + y)^n$$

The growth of the tiger can be therefore be modelled by the function:

$$h(t) = h_0(1 + y)^n$$

where $h(t)$ is the height at time t and n is the number of time units (in this case years). For the tiger:

$$h(t) = h_0(1 + 0.08)^n$$

or $$h(t) = 35(1.08)^n$$

This can be seen to be the equation of an exponential:

$$y = ta^x$$

where t *and* a *are constants,* x *is the independent variable and* y *is the dependent variable. After five years the tiger's height will therefore be:*

$$h(5) = 35(1.08)^5 = 35 \times 1.45 = 50.75 = 50.8\,\text{cm}$$

9.5.1.1 *Bacterial growth*

Bacteria divide by binary fission, so each cell produces two daughter cells. A bacterial population is characterised by a doubling time which measures how long it takes for the population to double, i.e. for the bacteria to divide. For *Escherichia coli* the doubling time is approximately 20 min but for *Bacillus subtilis* the doubling time is about 40 min. Bacterial growth will be exponential since after each doubling time the number of cells will increase by a factor of two so that the ratio between consecutive numbers is two. Suppose, therefore, that you have a population of N_0 cells and that this increases to a value of $N(t)$ after time t (min). If we follow the same argument as was given for increasing height in Example 9.5, we have:

$$N(t) = N_0 2^n$$

where n is the number of doubling times.

Example 9.6

Suppose we have one E. coli *cell and one* B. subtilis *cell in a culture medium. How many of each cell type will exist after 24* h?

We must first convert the time into generation or doubling times, so:

$$24\,\text{h} = 24 \times 60 = 1.44 \times 10^3\,\text{min}$$

The number of doubling times is therefore:

$$(1.44 \times 10^3)/20 = 72 \text{ for E. coli}$$

and

$$(1.44 \times 10^3)/40 = 36 \text{ for B. subtilis}$$

so using $N(t) = N_0 2^n$ *with* N_0 *assigned the value one we have:*

$$N(24) = 2^{72} = 4.72 \times 10^{21} \text{ for E. coli}$$

and

$$N(24) = 2^{36} = 6.87 \times 10^{10} \text{ for B. subtilis}$$

It can be seen that the faster generation time for *E. coli* could give it a competitive edge over *B. subtilis* since it is able to coloniso a nutrient-rich environment rapidly. In the above example *n* represents time in terms of the number of generations, but generation times vary between different bacterial strains and are also affected by the growth conditions. The equation for modelling bacterial growth would therefore be better if it was refined to give:

$$N(t) = N_0 2^{\alpha t}$$

In this case the exponent can be recorded in real time *t* (min) and the factor α acts a **growth constant**. It effectively takes into account the speed or rate of growth for the organism. One doubling would be given by:

$$\alpha t = 1$$

If the doubling time was 20 min you would have:

$$\alpha 20 = 1$$

$$\alpha = \frac{1}{20} = 0.05 \, \text{min}^{-1}$$

Hence α would be $0.05 \, \text{min}^{-1}$.

9.5.2 *Exponential decay*

Exponential decay processes are also common within biology; for example the loss of drug from a patient's blood by a combination of excretion, metabolism and sequestration into other biological compartments can be modelled using exponential decay. Loss of life due to an epidemic may lead to an exponential decay since if the population is large the infectious agent could be easily transmitted leading to high mortality, but as the population decreases in size so transmission decreases and the number of deaths decreases. In this section we will look at radioactive decay, since many experiments in life sciences use isotopes. The scientist can usually label small quantities of a biomolecule with a radioactive isotope, thus allowing it to be detected in a scintillation counter.

The unit of radioactive decay is the becquerel (Bq).

Example 9.7

One molecule of penicillin-binding protein 5 binds one molecule of penicillin. The protein is purified and incubated with 37 kBq of radioactively labelled penicillin (^{14}C-penicillin, at 35 MBq mmol^{-1}). After separation of the unbound penicillin by column chromatography, 10 kBq are found bound to the protein. Assuming all the protein present binds one molecule of penicillin per molecule protein, how much protein do I have?

The specific activity of the isotope was 35×10^6 Bq mmol^{-1}; 10×10^3 Bq *are bound so I have*:

$$((10 \times 10^3)/(35 \times 10^6))\,\text{mmol} = 2.85 \times 10^{-4}\,\text{mol}$$

or approximately 0.3 mmol.

^{14}C is very stable, but suppose we use ^{32}P instead. This has a half-life of 14.3 days, i.e. it decays to half its original activity in 14.3 days. If for some reason the experiment took four days the specific activity of the sample would have decreased and this would have to be considered. Exponential decay is modelled by the equation:

$$N(t) = N_0 2^{-\lambda t}$$

where $N(t)$ is the activity remaining at time t (days) and N_0 is the starting activity. λ is a decay constant which is unique to a given isotope. With bacterial growth we used growth constant α to convert the time into the number of doubling times, so α took growth rate into account. In this case the decay constant considers how long it takes for the sample to decay or break down. For example, we know that ^{32}P takes 14.3 days to decay to half its original activity, so:

$$N(t)/N_0 = 0.5 = 2^{-\lambda 14.3}$$

This can be solved using logarithms, as will be shown in Chapter 10, or we can plot $y = 0.5$ and $y = 2^{-x}$ to find where they intercept and thus discover x, where $x = \lambda \cdot 14.3$.

From Figure 9.5 the lines intercept when $0.5 = 2^{-x}$ and at this point $x = 1$. To find λ for ^{32}P we have:

$$1 = \lambda 14.3 \quad \text{so} \quad \lambda = 1/14.3$$

Table 9.5

x	-1	0	1	2	3
$y = 2^{-x}$	2	1	0.5	0.25	0.125
$y = 0.5$	0.5	0.5	0.5	0.5	0.5

Figure 9.5

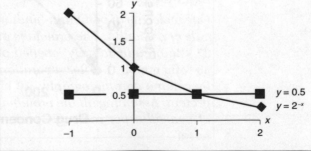

or, more generally,

$$\lambda = 1/\tau_{1/2}$$

where $\tau_{1/2}$ represents the half-life. This is therefore very similar to the exponential growth of bacteria, where the growth constant α represented 1/doubling time.

9.5.3 Geometric series

On occasions a life scientist will be performing an experiment where the output is linked to the input by an exponential function. In this case it is not always useful simply to take readings at equal intervals and it may be better to prepare your samples according to a geometric series. For example, if you are testing a drug against a population, be it cells or animals, the effect or response to the drug will not be linear. If you start with a drug concentration of 20 nM, then it would be better to increase this concentration geometrically. Consider the plots in Figures 9.6 and 9.7, where response is measured as the percentage of cells killed (from data in Tables 9.6 and 9.7).

You can see that with five points the geometric series covers the whole range of the response curve. In addition, at the start, where the readings are small and prone to error, we have more points so this will allow the curve to be plotted more accurately. With the linear scale we would need three more points to cover the full range of the curve

Table 9.6 Drug testing with a linear sequence of concentrations

Drug concentration/nM	20	120	220	320	420
Response/%	9	43	67	78	84

Figure 9.6
Dose response curve with linear sequence of concentrations.

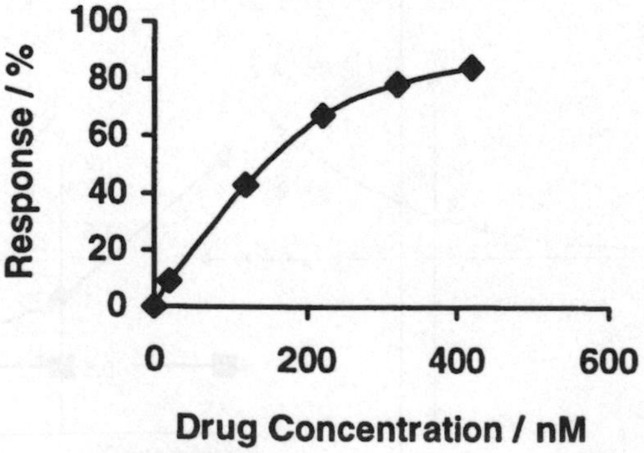

Table 9.7   Drug testing with a geometric series of concentrations

Drug concentration/nM	20	50	125	312	781
Response/%	9	20	47	76	92

Figure 9.7
Dose response curve with geometric sequence of concentrations.

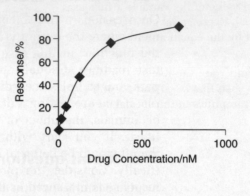

There are occasions where it is better to take readings according to a geometric sequence rather than a linear sequence

and between 0 and 100 nM we have only one point at 20 nM, compared with the 20 and 50 nM points in the geometric series.

So it can be seen that there may be occasions, such as in the case of toxicity testing, where using a geometric series to calculate the drug concentrations to be tested would be valuable.

Figure 9.8

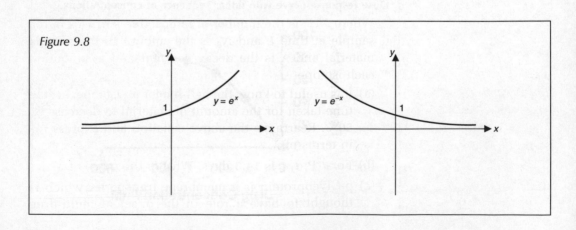

Summary

In biology many process increase or decrease exponentially. This form of plot tends to $y = 0$ but never actually reaches this line; hence $y = 0$ is said to be an asymptote. An exponential graph will very rapidly increase (or decrease) as the numerical value of the variable x increases (Figure 9.8). Exponential growth is described by the equation:

$$y = ta^x$$

where x and y are variables and a and t are constants such that $a > 0$ and $t > 0$. If $t = 1$, then this simplifies to the special case where:

$$y = a^x$$

Exponential decay would be described by a closely related equation:

$$y = ta^{-x}$$

These equations can be solved graphically or by using a logarithmic transformation as described in Chapter 10. If two variables x and y have an exponential realationship, it may be advantageous to examine values of y for a geometric rather than a linear sequence of x.

End of unit questions

1. If a bacterial culture contains $N(t)$ bacteria at time t, then the growth of the population can be modelled by the equation:

$$N(t) = N_0 \cdot 2^n$$

where N_0 is the number of bacteria at the start and n is the number of generation times that have occurred. For example, a *Bacillus subtilis* bacterium divides approximately every 40 min and a culture was found to contain 103 bacterial cells. How many cells are present after 10 h?

2. A radio-isotope decays to a non-radioactive form and the process follows an exponential decay curve. The equation to describe this process is given by:

$$N(t) = N_0 e^{-\lambda t}$$

Where $N(t)$ is the number of radioactive particles in the sample at time t, and N_0 is the amount in the starting material and λ is the decay constant and is unique to each isotope.

(a) It is useful to know the half-life for an isotope, i.e. the time taken for the amount of material to decrease by 50%. Rearrange the above equation and express $t_{1/2}$ in terms of λ.

(b) For ^{32}P, $t_{1/2}$ is 14.3 days. What is the value of λ?

(c) p-Glycoprotein is a membrane transporter which is thought to have a role in the onset of multi-drug resistance during cancer chemotherapy. It undergoes phosphorylation within the cell. In an experiment, p-Glycoprotein is detected using radioactive ^{32}P. When purchased, the ^{32}P for the labelling experiment had a specific activity of 6735 BqμM^{-1}. The labelling experiment was performed four days later and 2367 Bq of activity was incorporated.

  (i) Allowing for decay, what was the specific activity of the ^{32}P?

  (ii) If one molecule of ^{32}P binds to one molecule of p-Glycoprotein, what is the concentration of p-Glycoprotein?

3. A population increases at an annual rate of 4% to 360 000 over a period of ten years. Assuming exponential growth, what was the original size of the population?

4. A baby weighs 3.2 kg and one month later it weighs 3.4. kg. Assuming the weight increases according to a geometric sequence, what would the weight be four months after the first weighing?

5. The height of the baby in question 4 was also measured and was found to be 42 cm. The second measurement was taken two months later and found to be 48 cm.

  (a) Assuming that height increases according to an exponential monthly growth rate, what was the height one month after the first measurement?

  (b) What would the height be six months after the first measurement?

6. A drug is to be tested in tissue culture. It is expected that the dose response will be described by an exponential function. The first addition of drug gives a final concentration of 5 μM and the second addition gives 7.5 μM. What would you choose as your next five drug concentrations?

7. A drug is administered intravenously. The original blood plasma concentration is C_0 and the plasma concentration at time t (min) is C_p. That fraction of the drug which is eliminated per unit time is K (min^{-1}). For example, $K = 0.02$ min^{-1} implies that 2% of the drug is eliminated every minute. Elimination from the plasma will be due to metabolism, secretion and uptake. The concentration of drug at any given time is:

$$C_p = C_0 e^{-Kt}$$

The drug concentration decreases by 5.6% in 1 h. Find K.

10 Logarithmic Functions

10.1 Introduction

Logarithms (logs) are widely used within science, yet many students view them with trepidation. They have a range of functions and can be used to solve equations and linearise exponential functions. The most common use you are likely to encounter is to alter scales or to transform functions. In this chapter logarithms will be introduced and some examples of their use will be given. The objectives are:

(a) to introduce logarithms and develop confidence in their use;

(b) to show the relationship between logarithms and exponential functions;

(c) to discuss rules for manipulating logarithms;

(d) to show logarithmic transformations of power and exponential functions;

(e) to introduce semi-logarithmic plots and log–log plots.

10.2 Defining logarithms

Exponents were introduced in Chapter 5 with respect to power expressions. In Example 10.1 one thousand is represented as a power expression. The base used is ten and the exponent is three.

Example 10.1

$$1000 = 10^3$$

A **logarithm** (or log) is closely related to a power and can be used to write Equation (10.1) in a different manner, as shown in Example 10.2.

Example 10.2

$$\log_{10} 1000 = 3$$

This can be read as: 'The logarithm to the **base** ten of one thousand is equal to three'. In the case of logs, the base is represented by the subscript at the side of the log term. Notice that the value of $\log_{10} 1000$ is equal to the exponent in Example 10.1. This is because in both Examples 10.1 and 10.2 the base used is ten. Example 10.1 shows that one thousand can be expressed as base ten to the power three. Example 10.2 can be thought of as saying that if one thousand were expressed to the base ten, then its exponent would be three.

If a number is expressed to the base a then the log (to the base a) of that number is simply the exponent in the power term. Although this sounds complex, it is in fact relatively straightforward, as shown in Box 10.1.

Box 10.1

If	$y = a^x$
Then	$\log_a y = x$

This is illustrated in Example 10.3.

Example 10.3

$$10 = 10^1$$

so

$$\log_{10} 10 = 1$$

Worked examples 10.1

Express the following values in the form of $\log_{10}$:
(i) 100   (ii) 10^7 (iii) 1   (iv) $10^{2.3}$

Notice from Box 10.1 that if $\log_{10} y = x$ we must have a value such that $y = 10^x$. From Chapter 5 it is known that x can be positive or negative, but in *all cases* the value 10^x is positive and greater than zero (Chapter 9). This means that we can never find the log of y if y is negative or equal to zero, since we have said that y is to equal 10^x and 10^x is always greater than zero (Box 10.2).

Logarithms only exist for positive numbers

Box 10.2

If $\log_a y = x$ then $y = a^x$
Since $a^x > 0$, $y > 0$
so $\log_a y = x$ only if $y \geq 0$

It therefore follows that you cannot calculate the log of zero, nor can you take logs of negative numbers. Each positive number does have a logarithm and the original number is termed the antilogarithm (Example 10.4).

Example 10.4

The logarithm to the base ten of 100 *is equal to* 2

i.e. $\log_{10} 100 = 2$

The antilogarithm to the base ten of 2 *is equal to* 100

i.e. antilog$_{10}$ 2 $= 100$

It can be seen in Example 10.4 that to find the antilog you are forming a power term in which the number under investigation becomes the exponent: to calculate antilog$_{10}$ 2 you simply find 10^2. On your calculator you will find the 'log' key usually also has the exponential '10^x' as a second function. This is because they are **inverses** (Chapter 7). If you take the $\log_{10}$ of a number and then raise your answer to the base ten you get the original value back, as in Example 10.5.

Logarithms to the base *a* form the inverse to exponentials to the base *a*

Example 10.5

$$\log_{10} 10\,000 = 4$$
$$\text{antilog}_{10} 4 = 10^4 = 10\,000$$

In the same way as power terms can have different bases, logs can be calculated to different bases so long as the base is greater than zero (Example 10.6).

Example 10.6

$$25 = 5 \times 5 = 5^2$$
so
$$\log_5 25 = 2$$

It is therefore necessary to specify the base you are working in by writing it as a subscript, as shown above. In the life sciences there are really only three bases that are commonly used: ten, two and *e*.

10.2.1 *Logarithms to the base ten* ($\log_{10}$).

This base is widely used and will be found on a calculator as $\log_{10}$. Logs to the base ten are also called **common logs** and since they are the most widely used form of logarithm it is often written as 'log' without the subscript. Any log without a subscript is therefore assumed to be to the base ten.

10.2.2 *Logarithms to the base two* $(\log_2)$

This is not widely used, but may be applied in cases where a quantity alters in jumps of two. For example, bacterial growth occurs by binary fission (i.e. a cell splits into half to give two offspring). This process can be described by doubling times and modelled using $\log_2$.

10.1.3 *Natural logarithms* $(\log_e)$

Natural logs are calculated to the **base e** which can be approximated by the number 2.718. They are also called **Napierian logs** and are often written as 'ln' rather than '$\log_e$'. These logs are used to describe naturally occurring exponential processes and are related to common logs as shown in Box 10.3.

Box 10.3

$$\ln x = 2.303 \log_{10} x$$

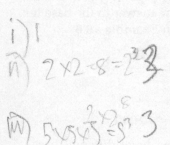

Worked examples 10.2

Evaluate the following without the use of a calculator:
(i) $\log_{10} 10$ (ii) $\log_2 8$ (iii) $\log_5 125$ (iv) $\log_4 16$

Suppose that you find yourself in a position where you need to calculate an unusual log, for example log to the base seven. This is not present on the calculator and the easiest way to find the answer is by using the equation given in Box 10.4. This is highlighted in Example 10.7.

Box 10.4

$$\log_a x = \frac{\log_{10} x}{\log_{10} a}$$

Example 10.7

Find $\log_7 30$.

$$\log_7 30 = \log_{10} 30 / \log_{10} 7 \quad \text{(Box 10.4)}$$
$$= 1.477/0.845$$
$$= 1.748$$

Notice that from Box 10.4 it is possible to show that $\log_a a = 1$, since

$$\log_a a = \frac{\log_{10} a}{\log_{10} a} = 1$$

This agrees with what we have said previously since we know that

$$10 = 10^1$$

and therefore $\log_{10} 10 = 1$ (Box 10.1)

The equation in Box 10.4 can also be used to highlight the fact that $\log_a 1$ always equals zero.

$$\log_a 1 = \log_{10} 1 / \log_{10} a \quad (Box\ 10.4)$$
$$= 0 / \log_{10} a \quad\quad (log_{10}\ 1 = 0)$$
$$= 0 \quad\quad\quad\quad (zero\ divided\ by\ anything\ equals$$
$$zero)$$

The two rules described above can be very useful when simplifying equations and are listed in Box 10.5.

Box 10.5

$$\log_a a = 1$$
$$\log_a 1 = 0$$

10.3 Rules for manipulating logarithmic expressions

There are three laws for the manipulation of logs which hold for any expression as long as all the logs being manipulated have the same base.

10.3.1 *Law for the addition of logarithms*

This law simply shows that if you are adding the logs of two numbers x and y and the logs have the same base, then the sum is equal to the log of the product xy (Box 10.6). Example 10.8 shows two routes to the same answer.

Box 10.6

$$\log_a x + \log_a y = \log_a(xy)$$

Example 10.8

$$\log_{10} 6 + \log_{10} 5 = \log_{10}(5 \times 6) \quad or$$
$$= \log_{10} 30$$
$$= 1.477$$

or

$$\log_{10} 6 + \log_{10} 5 = 0.778 + 0.699$$
$$= 1.477$$

10.3.2 *Law for the subtraction of logarithms*

This law states that if you are subtracting the logs of two numbers x and y and the logs have the same base, then the subtraction is equal to the log of the quotient x/y (Box 10.7 and Example 10.9).

Box 10.7

$$\log_a x - \log_a y = \log_a \left(\frac{x}{y}\right)$$

Example 10.9

$$\log_{10} 32 - \log_{10} 4 = \log_{10} \left(\frac{32}{4}\right)$$
$$= \log_{10} 8$$
$$= 0.903$$

or

$$\log_{10} 32 - \log_{10} 4 = 1.505 - 0.62$$
$$= 0.903$$

It is worth noting from this example that

$$\log_a 1 - \log_a y = \log_a \left(\frac{1}{y}\right) \qquad \text{(Box 10.7)}$$

But $\log_a 1 = 0$ \qquad\qquad\qquad (Box 10.5)

so $0 - \log_a y = -\log_a y = \log_a \left(\frac{1}{y}\right)$

Box 10.8

$$-\log_a y = \log_a(1/y)$$

10.3.3 Law for logarithms of power terms

This rule shows that multiplying the log of x by a value n is the same as calculating the log of a power where x is the base and n is the exponent (Box 10.9 and Example 10.10).

Box 10.9

$$n\log_a x = \log_a(x^n)$$

Example 10.10

$$10\log_{10} 10 = \log_{10}(10^{10}) \quad or \quad 10\log_{10} 10 = 10 \times 1$$
$$= 10 \qquad\qquad\qquad\qquad = 10$$

The rule shown in Box 10.9 can be used to simplify the log of a root, since roots can be represented as fractional indices (Chapter 5). This is illustrated in Example 10.11, and the general equation is given in Box 10.10.

Example 10.11

$$\sqrt[n]{a} = a^{1/n}$$

For example, the square root of ten is given by: $\sqrt{10} = 10^{1/2}$

so
$$\log_{10}(\sqrt{10}) = \log_{10}(10^{1/2})$$
$$= \tfrac{1}{2}\log_{10} 10$$
$$= \tfrac{1}{2} \times 1 = \tfrac{1}{2}$$

Box 10.10

$$\log_a(\sqrt[n]{x}) = \frac{1}{n}\log_a x$$

One occasion in which the formula in Box 10.10 may be useful is when you need to calculate an unusual root, for example $\sqrt[5]{26}$. If you calculator has an $x^{1/y}$ button, you can enter this directly as '$26x^{1/y}5 =$' but if you do not have this function you can use logs as shown in Example 10.12.

Example 10.12

$$\sqrt[5]{26} = 26^{1/5}$$

so
$$\log_{10}(\sqrt[5]{26}) = \log_{10}(26^{1/5})$$
$$= \tfrac{1}{5}\log_{10} 26$$
$$= 0.2 \times 1.415 = 0.283$$

Since
$$\text{antilog}_{10}(\log_{10}(\sqrt[5]{26})) = \text{antilog}_{10} 0.283$$
$$\sqrt[5]{26} = 10^{0.283}$$
$$= 1.919$$

Worked examples 10.3

(a)   Simplify the following:
(i) $\log_{10} 2 + \log_{10} 6$ (ii) $3\log_{10} 2 - 2\log_{10} 4$
(iii) $2\log_{10} a - \log_{10} 6$

(b)   If $\log_{10} 6 = 0.78$ and $\log_{10} 2 = 0.30$, calculate the following without a calculator:
(i) $\log 2^6$ (ii) $\log 12$ (iii) $\log 36$ (iii) $\log 3$

10.4 Using logarithms to transform data

A log is the inverse function of an exponential, assuming that the base is the same in both cases. This is shown mathematically in Example 10.13.

Example 10.13

Logarithms can be used
to solve equations
containing indices

$$f(x) = \log_e x$$
$$g(x) = e^x \qquad \text{and } \log_e(e^x) = x$$

so
$$f(g(x)) = f(e^x)$$
$$= \log_e(e^x)$$
$$= x$$

Logs can therefore be used to help solve exponential equations such as Example 10.14.

Example 10.14

$$10^x = 14 \qquad \text{Taking logs to the base 10,}$$

$$x = \log_{10} 14 = 1.15$$

The reverse is also true, since logs can be solved by the use of exponentials (Example 10.15).

Example 10.15

$$\log_{10} x = 2.97$$

Therefore $$x = 10^{2.97}$$
$$= 933.25$$

Logarithms can also be used to transform both exponential functions and power functions into straight-line forms which can then be plotted.

10.4.1 *Logarithmic transformation of exponential functions*

Many biological process are exponential, yet exponential equations are not very user-friendly. In most situations it is easier to transform the data into a straight-line form and use the transformed equation to analyse the data.

An exponential function is defined by the general equation:

$$y = ta^x$$

where y and x are variables and a and t are constants. Using log rules, this can undergo a logarithmic transformation to give:

$$\log_{10} y = \log_{10} t + \log_{10} a^x$$
$$= \log_{10} t + x \log_{10} a$$

If we define a new variable Y where $Y = \log_{10} y$ and we create two constants, $A = \log_{10} a$ and $T = \log_{10} t$, then the above equation can be expressed in the form:

$$Y = T + xA$$

which can be rearranged to give

$$Y = Ax + T.$$

This form of equation describes a linear function of the form:

$$y = mx + c$$

If Y (i.e. $\log_{10} y$) is plotted against x, then the gradient of the line would be A (i.e. $\log_{10} a$) and the intercept would be T (i.e. $\log_{10} t$) as described in Chapter 7. This form of plot is termed a semi-logarithmic plot and is discussed in Section 10.5.

10.4.2 *Logarithmic transformation of power functions*

Remember that a power function is not the same as an exponential function. In the case of a power function the exponent is a constant, whereas for an exponential function the exponent is the variable x (Chapter 8). Consider the function

$$y = ax^n$$

where x and y are variables and a and n are constants. If x and y are greater than zero, we can apply a logarithmic transformation to give:

$$\log_{10} y = \log_{10} a + n \log_{10} x$$

Forming new variables $Y = \log_{10} y$ and $X = \log_{10} x$ and the constant $A = \log_{10} a$, we can substitute these into the above equation to give:

$$Y = A + nX$$

which when rearranged gives the linear equation

$$Y = nX + A$$

Once again this is the equation of a straight line, and if X (i.e. $\log_{10} x$) is plotted against Y (i.e. $\log_{10} y$) this will produce a line with gradient n and the intercept A (i.e. $\log_{10} a$). This is termed a log–log or double-logarithmic plot and is described in section 10.6.

10.5 Semi-logarithmic plots

When either the x- or y-axis of a plot is given a logarithmic scale, the coordinate system is said to be semi-logarithmic. If one of the variables spans a very large range, for example 1 to 10 000 it is hard to prepare a meaningful plot, but if this scale is plotted logarithmicaly the range would be condensed to give a scale from 0 to 4 since:

$$\log_{10} 10 = 1 \quad \text{and} \quad \log_{10} 10\,000 = 4$$

This form of plot is widely used within the life sciences. It is especially common within toxicology, since when looking at the response of a cell or an organism to a drug or toxic agent a wide concentration range may be used, and it is the log of the dose that is biologically important. This is illustrated below. In Figure 10.1 the dose (in mM, for example) is plotted against the percentage response. This could be the number of cells killed. It can be seen that there is initially a rapid response which then appears to level out, but it is hard to determine any detail because the scale covers a wide range.

The same plot is shown in Figure 10.2, but with a log scale for the dose. It can be seen that this gives a sigmoidal curve

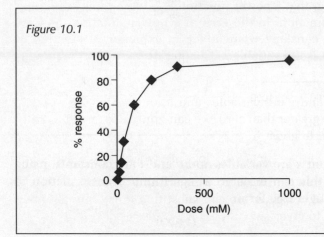

Figure 10.1

Dose (mM)

Source: Data are fictitious with the dose being measured in arbitrary units and the % response indicating the percentage of the sample population killed by the agent.

and we can now observe the biologically relevant detail. For example, there appears to be little or no toxicity below four dose units; hence there maybe a threshold level below which there is no observed effect (NOEL). NOEL is important in toxicology for setting exposure limits. We can also see over what range increasing dose causes increasing response, so this is a much clearer and relevant way of presenting toxicity data.

Notice that in Figure 10.1 a value is plotted for dose 0 M. We therefore have a problem since $\log 0$ is undefined, yet if you are studying the effect of a drug you must have a control containing no drug, i.e. 0 M. The logarithmic transformation is what is biologically relevant, so you need to deal with this $x = 0$ value. The method chosen is usually to transform the data using the equation

$$X = \log(x + 1)$$

so that you can then plot X against y. This transformation has been used to produce Figure 10.2.

10.5.1 *Exponential functions*

If two variables are related by an exponential function:

$$y = ta^x$$

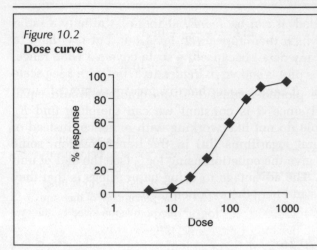

Figure 10.2
Dose curve

Source: Data are fictitious with the dose being measured in arbitrary units and the % response indicating the percentage of the sample population killed by the agent.

where y and x are variables and a and t are constants, using log rules this can undergo a logarithmic transformation to give a straight-line form:

$$Y = T + xA$$

where $Y = \log y$, $T = \log t$ and $A = \log a$.

A semi-logarithmic plot can therefore be obtained by plotting $\log y$ against x. An example of this is given below.

10.5.1.1 The Arrhenius equation

To undergo a chemical reaction, molecules must overcome an energy hill which is termed the activation energy (E_a). If the reacting species come together and have energy equal to or greater than E_a, they will react to form product. If they come together but have less energy than E_a, they will remain as substrates and separate. If you heat up the system, the molecules in it gain energy so that more molecules have enough energy to react and so the reaction proceeds at a faster rate. The relationship between the reaction's rate constant (k) and the activation energy is given by the Arrhenius equation:

$$k = Ae^{-E_a/RT}$$

This is an exponential function in which A is a constant for a particular reaction, T is the temperature measured in kelvin and R is the gas constant ($8.314\,\mathrm{J\,K^{-1}\,mol^{-1}}$). The independent variable is therefore T (K), which the scientist can control. The dependent variable is k, the rate constant which of course depends on T and is the value which is being measured at different temperatures. This equation can therefore be transformed with logs as follows:

$$\ln k = -\frac{E_a}{RT} + \ln A$$

$$= -\frac{E_a}{R} \times \frac{1}{T} + \ln A$$

If $\ln k$ was plotted against $1/T$ the gradient would equal $-E_a/R$ and since R is constant we can therefore find E_a. Some people do not like working with e; hence, instead of using natural logarithms (ln) in the transformation, some books will give the equation using $\log_{10}$ (see the end of unit questions). The advantage to using natural logs is that they have base e and here the exponential has base e:

$$\log_e e^x = x$$

Example 10.16

The simplest way to obtain E_a experimentally is to measure the maximum velocity the reaction can obtain at each temperature ($V_{max}/mM^{-1} min^{-1}$) and plot $\log(V_{max})$ against $1/T$. Some typical data are gathered in Table 10.1.

Table 10.1 Effect of temperature on enzyme activity

T/K	$(1/T) \times 10^3/K^{-1}$	$V_{max}/$ $mM^{-1} min^{-1}$	$\log(V max)$
293	3.413	4.5	0.653
303	3.300	8.65	0.937
308	3.247	11.8	1.071
313	3.195	15.96	1.203
318	3.145	21.36	1.33

Source: Modified from data on the hydrolysis of lactose by β-galactosidase, in *Biochemical Calculations*, 2nd edn, 1976, I.H. Segal © John Wiley & Sons (1976). Reprinted by permission of John Wiley & Sons Ltd.

We can now plot $1/T$ against $\log(V_{max})$ on a normal piece of graph paper as shown in Figure 10.3.

We have now plotted $\log(V_{max})$ on a linear scale. It is worth mentioning, though, that semi-logarithmic graph paper can be obtained, with a log scale incorporated into it so that you could simply plot $1/T$ against V_{max} with the V_{max} values being plotted on the log scale to give the logarithmic transformation. This form of graph paper is useful since it eliminates the need for taking logs of the data; but you have to remember that in the first 'block' each line represents one unit, in the second ten units, in the third a hundred units and so on. It can therefore be difficult to plot the data accurately. A graph in which V_{max} is plotted on a log scale is shown in Figure 10.4.

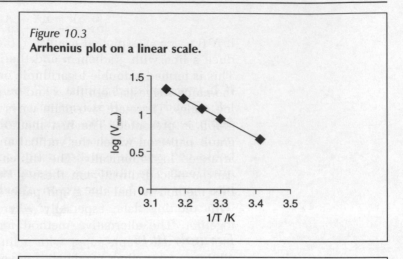

Figure 10.3
Arrhenius plot on a linear scale.

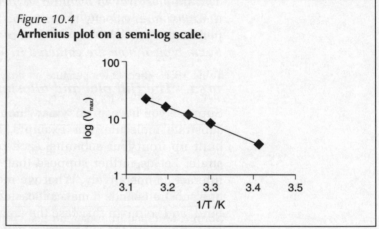

Figure 10.4
Arrhenius plot on a semi-log scale.

Care should be taken if using computer packages that the correct plot has been prepared, i.e. that you have either plotted $\log x$ on a linear scale, or x on a log scale, but not $\log x$ on a log scale!

10.6 Double-logarithmic plots

Consider the equation:

$$y = ax^n$$

If x and y are greater than zero we can apply a logarithmic transformation to give:

$$\log y = \log a + n \log x$$

We can substitute new variables $Y = \log y$ and $X = \log x$ and the constant $A = \log a$ into the above equation to give the general linear equation:

$$Y = nX + A$$

If X (i.e. $\log x$) is plotted against Y (i.e. $\log y$), this will produce a line with gradient n and the intercept A (i.e. $\log a$). This is termed a double-logarithmic or a log–log plot and, as the name suggests, both the x and y variables are plotted on log scales. There are two main ways in which this form of graph is presented. The first method involves the use of graph paper on which the vertical and horizontal lines are arranged logarithmically. The values for variables x and y can be placed directly on these scales. The problem with this method is that the graph paper can lead to problems with plotting data, especially where the lines are close together. The alternative method for preparing a log–log plot is to use graph paper with a linear scale. In this case all the x and y values need to be converted to variables X and Y where $X = \log x$ and $Y = \log y$. An example of a log–log plot is given in Section 10.6.1.

10.6.1. *The Hill plot and allosteric enzymes*

Suppose you have an enzyme which can bind more than one substrate molecule. For example, suppose the enzyme is built up from four subunits, each of which binds one substrate. Let us further suppose that these four binding sites interact co-operatively. What we mean by this is that when one substrate binds it makes it easier for the next substrate to bind, and so on. In this case the enzyme may initially have a low affinity for the substrate so as we increase the substrate concentration there is little activity because interaction between the substrate and the enzyme is limited. Eventually we have a high enough concentration of substrate for a single molecule to bind to the enzyme. This binding affects all the other sites, making their affinity for the substrate increase. Because the enzyme now has a higher affinity for the substrate, the next molecule of substrate will quickly bind. This increases the enzyme's affinity for the substrate even further, so the next molecule of substrate is picked up almost immediately. In this way we have gone from little activity to high activity very rapidly. This is described by the simple sequential interaction model of allosterism, and allosteric enzymes of this type are essential within the cell since they act as switches in metabolism. In response to changes in substrate concentration, they rapidly increase or decrease activity turning metabolic pathways on and off. The equation describing the kinetics of a multi-site enzyme-based reaction is termed the Hill equation:

$$v = \frac{V_{\max}[S]^n}{K' + [S]^n}$$

where n is the number of binding sites, v is the initial velocity measured at a given substrate concentration [S], $V_{\max}$ is the maximum velocity which can be obtained under these conditions and K' is a constant. The graph of this function is sigmoidal, as shown in Figure 10.5.

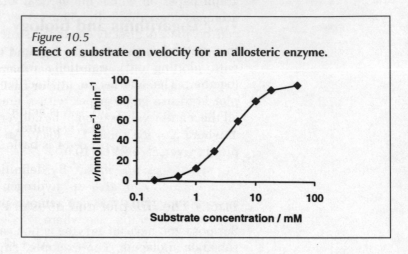

Figure 10.5
Effect of substrate on velocity for an allosteric enzyme.

The Hill equation is a power function, so it can be converted into a straight-line form using logarithmic transformations:

$$\frac{v}{V_{\max}} = \frac{[S]^n}{(K' + [S]^n)}$$

$$V_{\max}[S]^n = vK' + v[S]^n$$

$$V_{\max}[S]^n - v[S]^n = vK'$$

$$[S](V_{\max} - v) = vK'$$

$$\frac{[S]^n(V_{\max} - v)}{v} = K'$$

$$n\log[S] + \log\frac{(V_{\max} - v)}{v} = \log K'$$

or $$\log\frac{(V_{\max} - v)}{v} = \log K' - n\log[S]$$

$$= -n\log[S] + \log K'$$

implying that $$\log\frac{v}{(V_{\max} - v)} = n\log[S] - \log K'$$

So we have a straight-line equation with

$$Y = \log \frac{(V_{max} - v)}{v} \quad \text{and} \quad X = \log[S].$$

The gradient of this line is n, the number of binding sites. An example of this plot is given in the end of unit questions.

10.7 Logarithms and biology

Within biology we often use logs and they are a very important tool. You will certainly meet them when you study pH. pH is a measure of how acidic or basic a solution is:

pH 1–6 is acidic
pH 7 is neutral
pH 8–14 is basic

But what does this mean? By definition an acid is a compound which can 'give up' hydrogen ions (protons) and a base is something that can remove protons from solution. For example, in the case where

$$HA \leftrightarrow A^- + H^+$$

HA is an acid since it gives up the proton, H^+. Upon dissociation HA produces A^- and since A^- can remove protons from solution to form HA, it must be a base. Since A^- is formed from HA it is said to be the **conjugate base** of HA. The more readily HA releases protons the stronger the acid. For example, hydrochloric acid (HCl) is a **strong** acid and can be assumed to ionise fully so that 0.1 M acid produces 0.1 M protons:

$$HCl \rightarrow H^+ + Cl^-$$

Formic acid (HCOOH) is a weak acid and only some of the acid molecules release their protons, so 0.1 M acid will not dissociate to produce 0.1 M protons. A solution of 0.1 M formic acid therefore has a lower acidity than a 0.1 M solution of hydrochloric acid. When pH is measured it is 'the number of hydrogen ions present' which is being recorded. So measuring the pH indicates how strong the acid or base is.

Example 10.17

Suppose that you are measuring the acidity (proton concentration) in five solutions and that the range of hydrogen ion concentrations found covers the pH range 1–5. The data are given in Table 10.2 and illustrated in Figure 10.6 and 10.7.

Table 10.2   Variation of hydrogen ion concentration with pH

Sample number	pH	Hydrogen ion concentration/M
1	1	0.1
2	2	0.01
3	3	0.001
4	4	0.0001
5	5	0.000001

Figure 10.6
Plot of sample number against proton concentration.

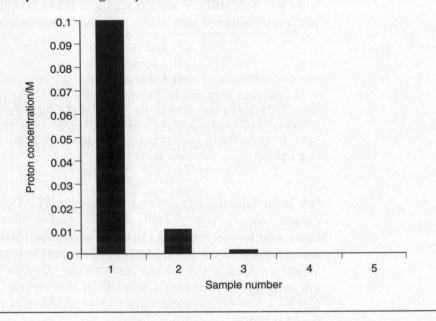

You can see from Figure 10.6 that if acidity is measured directly in terms of proton concentration it is impossible to distinguish between the hydrogen ion concentrations, and therefore the acidity, of samples 3, 4 and 5.

This is because in this example the hydrogen ion concentration has changed by five orders of magnitude. The scale in Figure 10.6 therefore covers too great a range to allow the small values to be distinguished. Suppose we plot the pH of the samples as in Figure 10.7. In this case acidity is measured in terms of pH and you can clearly distinguish between the acidities of all the samples.

Considering that the full acidity range covers pH 1–14 (proton concentrations 0.1 M to 0.000 000 000 000 01 M) it should be obvious that if you were measuring the hydrogen

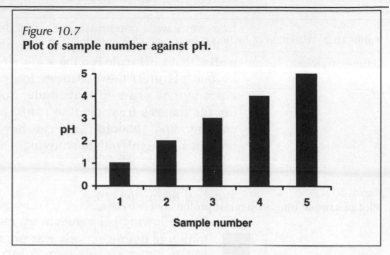

Figure 10.7
Plot of sample number against pH.

ion concentrations present, the scale would have to change by 14 orders of magnitude. In contrast, the pH scale varies by only one order of magnitude. So how are pH values related to hydrogen ion concentrations? If we write out the hydrogen ion concentration in terms of powers of 10, we have

$$pH\,1 = 10^{-1}\,mol\,litre^{-1}$$
$$pH\,3 = 10^{-3}\,mol\,litre^{-1}$$
$$pH\,5 = 10^{-5}\,mol\,litre^{-1}$$

Now it can be seen that we can quite easily deal with these concentrations if we use logs (Example 10.17).

Example 10.17

$$\log_{10}[H^+] = \log_{10}[0.1] \quad \text{where } [] = \text{concentration in mol\,litre}^{-1}$$
$$= \log_{10}[10^{-1}]$$
$$= -1$$

It easier to use positive numbers rather than negative numbers, so we define pH as given in Box 10.11.

Box 10.11
$$pH = -\log_{10}[H^+]$$
or
$$pH = \log_{10}(1/[H^+])$$

Notice that in this case logarithms are being used to condense a scale which covers many orders of magnitude. This logarithmic transformation of the data is a common means of condensing a scale and is especially relevant in toxicity studies, where a drug may be tested against a cell line over a wide concentration range. In this case, it is the log of the dose that has biological relevance. It is important to realise that a pH scale is a log scale, so that if the pH changes by one pH unit, the hydrogen ion concentration actually alters by one order of magnitude. So, if in an experiment the pH changes from pH 8 to pH 6, you change the proton concentration 100-fold. This can have a vast effect on the biological system you are studying.

Worked examples 10.4

(a) The following pH values were recorded. What concentration of hydrogen ions was present?
   (i) 5.0   (ii) 7.4   (iii) 10.2   (iv) 2.9

(b) What would be the pH for the following hydrogen ion concentrations?
   (i) 0.001 M   (ii) 1.1×10^{-10} M   (iii) 10^{-4} M
   (iv) 7.8×10^{-8} M

Summary

Logarithms act as an inverse operation for exponential functions and, although they can have any base, they are often found to the base ten or the base e. The base e is especially relevant to biological systems since it is used to describe naturally occurring exponential functions. These two forms of log are related by the equation:
$$\ln x = 2.303 \log x$$
Logs to the base a can be calculated using the equation:
$$\log_a x = \frac{\log_{10} x}{\log_{10} a} \quad or \quad \log_a x = \frac{\ln x}{\ln a}$$
It is worth remembering that $\log_a a = 1$ and $\log_a 1 = 0$, since these can be used to simplify equations. Equations containing logs to the same base can be also simplified using the following rules:

(i) $\log_a x + \log_a y = \log_a(xy)$

(ii) $\log_a x - \log_a y = \log_a\left(\frac{x}{y}\right)$

(iii) $-\log_a y = \log_a\left(\frac{1}{y}\right)$

(iv) $n\log_a x = \log_a(x^n)$

Logs can also be used to help calculate unusual roots, since:
$$\log_a \sqrt[n]{x} = \frac{1}{n}\log_a x$$
Logs have many applications in biology and may be used to condense scales which cover many orders of magnitude such as in the case of pH, or they can be used to convert power and exponential functions into straight-line forms. The exponential function:
$$y = ta^x$$
Transforms to:
$$\log y = \log t + x\log a$$
Hence a semi-logarithmic plot of $\log y$ against x will give a straight line of gradient $\log a$ and y intercept $\log t$. If a graph appears exponential it is therefore worth trying to show the data on a semi-log plot.
The power function:
$$y = ax^n$$
transforms to:
$$\log y = \log a + n\log x$$
Hence a log-log plot of $\log y$ against $\log x$ will produce a gradient of n and a y intercept of $\log a$.

End of unit questions

1. Simplify the following equations.

    (a) $\log x + 5\log y$ (b) $2\log t - 4\log t$

    (c) $0.5\log((9m)^2)$ (d) $\log(a+b) + \log(a-b)$

2. Solve the following:

    (a) $\log 5x = 3.7$ (b) $\log(4m-3) = 0.9$

    (c) $\ln x = 1.8$ (d) $\log 2x + 3\log x = 2.2$

3. The body must maintain its blood plasma pH at pH 7.4. If this pH changes it can have severe effects on metabolic reactions and the skeleton. A patient is admitted to hospital with chronic kidney disease and impaired renal acid excretion, and could have developed chronic acidosis. If this is the case the blood will be more acidic than pH 7.4.

    (a) If the blood pH was normal, what should the patient's blood hydrogen ion concentration be in $mol\,litre^{-1}$?

    (b) It is found that the blood plasma contains a concentration of $6.3 \times 10^{-8}\,M$ of hydrogen ions.

        (i) Is the blood more acidic or more basic than it should be?

        (ii) What is the pH of the blood?

4. The Arrhenius equation (Section 10.5.1.1) is given by

$$k = Ae^{-E_a/RT}$$

    where A is a constant for a particular reaction, T is the temperature measured in kelvin and R is the gas constant $(8.314\,J\,K^{-1}\,mol^{-1})$. E_a is the activation energy and k is the rate constant for that partcular reaction. This can be transformed to a straight-line form by using a logarithmic transformation. Transform the equation using logs to the base 10.

5. The data in Table 10.3 were obtained for an enzyme-catalysed reaction. Using the straight-line form of the Hill equation, find the number of binding sites (n).

6. The level of ionisation of an acid and its conjugate base is related to the pH of the system, and the relationship is given by the Henderson–Hasselbach equation:

$$pH = pK_a + \log \frac{[A^-]}{[HA]}$$

    Most anaesthetics exist in two forms—a protonated, charged form and an uncharged form. It is the uncharged

Table 10.3

Substrate concentration/mM	Initial velocity/μmol litre^{-1} min^{-1}
0.62	1.54
0.12	5.90
2.5	20.00
5.0	50.00
10	80.00
40	95.50
80	99.6

form which is active since this can partition into membranes.

$$BH^+ \leftrightarrow B + H^+$$

An anaesthetic such as prilocaine has a pK_a of 7.7. The cell can be considered to be at about pH 7.4.

(a) What is the ratio of the charged form to the uncharged form?

(b) If prilocaine entered the gastric tract where the gastric juice is at pH 2, what would the new distribution of base (B) to conjugate acid (BH$^+$)?

(c) The effect of pH on the ionisation of the drug is an important consideration since this affects uptake. Would prilocaine be more or less effective at pH 2?

7. A drug is administered intravenously. The original blood plasma concentration is C_0 and the plasma concentration at time t (min) is C_p. That fraction of the drug which is eliminated per unit time is K (min^{-1}). For example, $K = 0.02$ min^{-1} implies that 2% of the drug is eliminated every minute. Elimination from the plasma will be due to metabolism, secretion and uptake. The concentration of drug at any given time is:

$$C_p = C_0 e^{-Kt}$$

From the data in Table 10.4, find the following: C_0, K and the time taken for the drug to drop to $C_0/2$.

Table 10.4

C_p/μM	t/min
70	1
57	2
47	3
34	4
24	5

11 Introduction to Statistics

11.1 Introduction

Within the physical sciences there are many problems which may have an exact answer, but in the life sciences many of the questions asked may not have a fixed answer. For example, how much does a three-week-old baby weigh? If you go to a local maternity ward and weigh a few three-week-old babies you will find that their weights vary, yet if, for example, you produce baby clothes you need to have an idea of how big a baby will be at the different stages of its life. In this case it would be logical to weigh a number of three-week-old babies and then use this data set to try and estimate what the 'average weight' of a baby would be at this age. The process of taking a few representative measurements and then trying to assign parameters to the whole group is termed statistics. There are a number of questions which students should consider. For example, what is meant by 'representative data' and how accurate is the average with which you are trying to describe the whole population? It is these and related questions that will be considered in this chapter, the aims of which are:

(a) to introduce the normal distribution;

(b) to discuss means, modes and medians as average measures of a population;

(c) to discuss sample variability and methods of measuring it with variance, standard deviation and standard error of the mean;

(d) to introduce the idea of confidence intervals and the *t*-distribution.

11.2 Sampling

Let us return to the question posed in the Introduction: what is the weight of a three-week-old baby? Obviously this will vary, but we can determine an 'average' measure of weight for three-week-old babies; the question is, what do we mean by average and how confident can we be that this will represent the true weight of the next three-week-old baby we meet? To answer these questions we need to understand something about the range of data values that are possible and the frequency with which any given weight occurs. This is termed the **distribution**.

Before this distribution can be studied, however, the original question involving the weight of three-week-old babies needs to be clarified. For example, do male and female babies have the same weight? Do babies from different ethnic origins or from different countries have comparable weights? What about breast-fed verses bottle-fed babies? This perhaps highlights how important it is to consider the question being asked, because in trying to find an answer to it we will have to take some measurements, and for these to be of use they must be representative of the population in which we are interested. It may be that questions such as these make us focus on the real problem; for example, we may realise that what we are interested in is actually three-week-old male babies, born in the UK. We will assume that the other parameters can be ignored for the purpose of this chapter.

It is obviously not possible to measure all the three-week-old baby boys in the UK, so we will measure the weight of a sample and use this to estimate the average weight of the group or **population**. Rather than take all our measurements at one hospital, it would be better to take ten measurements from around the country to limit any regional variation.

Suppose all ten readings were exactly 5 kg. In this case it can probably be assumed that the weight of a three-week-old British boy is 5 kg. If the sample size is increased to 100 and all the readings were still 5 kg, then it is even more certain that three-week-old boys are 5 kg in weight. In other words, the bigger the sample the more confident we would be that our estimate was correct, since it is based on a bigger data set. In reality, if we took ten readings they would be likely to vary so we would have to calculate the average.

At this stage it is worth while looking at the data: if any readings are very different from the rest, you should return to them and check that they are correct. If so, then they must remain; but there could be reasons for removing a data point—for example, you may have made an error in taking the reading in which case the measurement should be repeated, or the object being measured may not be representative of the sample in which you are interested. When measuring three-week-old babies you find nine of the ten readings are in the range of 3.5–5.5 kg but one value is recorded correctly as 0.6 kg. On investigation it is found

Ensure that you have clearly defined the investigation and that the data are representative of the population

The larger the sample size, the more accurately the population can be modelled that this baby was not carried to term but was born ten weeks prematurely. Do you include this point? It is in the later weeks of pregnancy that babies gain most weight, so the 0.6 kg reading is not really representative of babies which are term (usually taken to be 38–42 weeks). In this case it may be better to clarify the question—What is the weight of three-week-old male babies born in the UK after being taken to term during pregnancy? A different reading can then be taken to replace this non-representative value.

If we repeat the work and take another ten measurements, we will probably get a different average. Both of these estimates are correct and the bigger the sample the more realistic they will be, but how sure are we that they represent the whole population? To answer this question we need to consider how variable the data are, to measure this variability and to use this measure to inform us and other workers how representative our estimate is.

11.3 Normal distribution

Ten male babies from around the UK were weighed at three weeks and the frequency table was constructed (Table 11.1).

Table 11.1

Baby weight/kg	Frequency
0.0–0.9	0
1.0–1.9	0
2.0–2.9	1
3.0–3.9	2
4.0–4.9	5
5.0–5.9	2
6.0–6.9	0

Source: Based on centile charts provided by the Health Education Authority (1993). Reproduced with kind permission from the Health Education Authority.

This can be represented by the histogram shown in Figure 11.1.

You can see that most of the readings are clustering around a central value. If we were to increase the sample size this would become even more apparent. For example, suppose we take 100 measurements. In this case we will decrease the interval in the frequency table so that we can obtain a more accurate idea of the most common weight (Table 11.2).

The data are illustrated in the histogram in Figure 11.2.

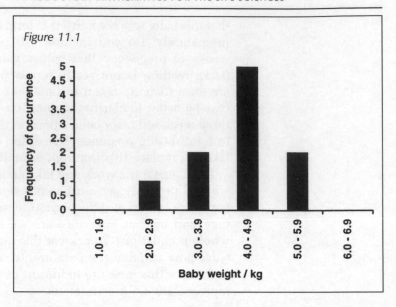

Figure 11.1

Table 11.2

Baby weight/kg	Frequency
2.5–2.9	0
3.0–3.4	8
3.5–3.9	20
4.0–4.4	37
4.5–4.9	21
5.0–5.4	11
5.5–5.9	2
6.0–6.4	0

Source: Based on centile charts provided by the Health Education Authority (1993). Reproduced with kind permission from the Health Education Authority.

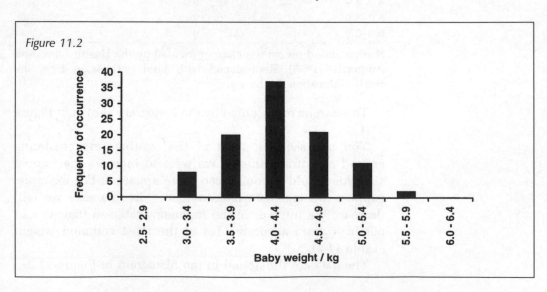

Figure 11.2

If we continued to increase the sample size and at the same time we kept decreasing the weight interval, we would end up with a smooth, bell-shaped curve as shown in Figure 11.3.

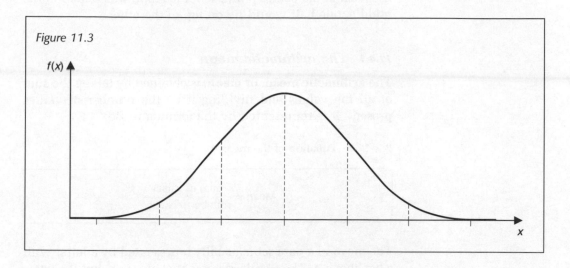

Figure 11.3

This bell-shaped curve is typical of a **normal distribution**; this form of distribution is obtained due to the natural variability in the sample. This would be the usual distribution used to approximate studies involving measures of weight, length and other forms of continuous measurement. The fact that data sets involving continuous measurement will eventually form a normal distribution as the sample size increases is known as the **central limit theorem**. This is beyond the scope of this text but can be found in most statistics books. The curve in Figure 11.3 shows the amount of variability present: the greater the spread, the greater the variability. Data sets with this form of distribution can be analysed by **parametric tests**. These tests make assumptions about the data, based on the normal distribution. Not all data sets are described by the normal distribution and if plotted some data give differently shaped curves instead of the bell-shaped curve Figure 11.3. Analysis of these data sets must involve **non-parametric tests.**

11.4 Means, medians and modes

If you have a data set and you want to describe the population as a whole, you need to assign a number which typifies the data. This kind of value is termed an average and exists in three main forms, those of the mean, median and mode. These will each be described in turn. The median and mean can also be used to gain insight into the symmetry of the distribution. If the distribution is normal, it will be symmetrical about the centre point: so if the graph was folded in the middle, one half would lie on top of the other.

11.4.1 *The arithmetic mean*

The **arithmetic mean**, or **mean**, is obtained by taking the sum of all the values and dividing it by the number of values present. It is represented by the formula in Box 11.1:

Box 11.1 **Equation of the mean.**

$$\text{Mean} = \frac{\text{sum of values}}{\text{number of values}}$$

The mean of a data set is usually represented by a letter with a bar above it. The standard letter of choice is x, but the letter may be defined with respect to the algebraic term you are using. For example, the length of a dachshund is denoted by the letter l (cm). Five dogs were measured and the mean length could therefore be denoted by $\bar{l}$. The bar above the letter denotes the mean. If a textbook is referring to the true mean of the population, and not an approximation calculated from a data set, the symbol μ is used. The sample size is usually denoted by the letter n. In textbooks you will often see the equation using a summation sign, as shown in Box 11.2.

Box 11.2 **Algebraic equation of the mean.**

$$\text{Mean} = \frac{\sum_{i=1}^{n} x_i}{n}$$

The equation in Box 11.2 can be read as the sum of all the data points x_i where $i \in \{1, 2, 3, \ldots, n\}$ divided by the number of data points, n.

If the data follows a normal distribution, the mean for the population will be the value that occurs at the centre of the curve. The main disadvantage with mean values is that they are strongly influenced by **outliers**. Outliers are single results which, if excluded from the calculation, would have a sig-

The mean occurs at the centre of a normal distribution

nificant effect on the result. As we discussed above, if a reading is erroneous or if it is known not to be representative of the data set then it may be removed, but if there is no obvious reason for the existence of the outlier then it must remain. To remove such values simply because they do not fit with what you expect should be frowned upon, since the removal of such points is not only poor science but is in fact fraudulent. The effect of outliers is highlighted in Example 11.1.

The mean is strongly influenced by outliers

Example 11.1

The weights of women in a class at a sixth-form college were measured and the data in Table 11.3 were obtained. The number of female students, n, was 11.

This is plotted in Figure 11.4; it can be seen that there is what may be an outlier to the right of the histogram. This was checked and found to be a valid reading. Therefore it should not be removed.

Let us consider the actual values recorded in the above example.

weights (kg) = {54.2, 56.0, 58.1, 59.3, 60.2, 60.7, 61.0, 62.2, 63.0, 64.6, 70.1}

If we calculate the mean, we find the mean weight for the group is 61.0 kg; if the last point was removed, the average would become 59.9 kg. It can be seen that the outlier has changed the mean by a considerable amount and this would obviously become even more significant if the outlier was further away from the main body of data or if the sample size decreased. In fact, assuming that the women in the group were of average height, the national mean would be expected to be about 62 kg and all of the weights measured would be considered normal.

If the data are given in a frequency table (Chapter 8) as in Example 11.1, then to calculate the mean you must multiply each data point by its frequency of occurrence. This is shown in Example 11.2.

Example 11.2

Nine herring were caught and the amount of vitamin D present in each was calculated per 100 g of herring. The results are shown in Table 11.4 and plotted in Figure 11.5. Find the mean.

Table 11.3

Weight/kg	Number of individuals
54–55	1
56–57	1
58–59	2
60–61	3
62–63	2
64–65	1
66–67	0
68–69	0
70–71	1

Figure 11.4

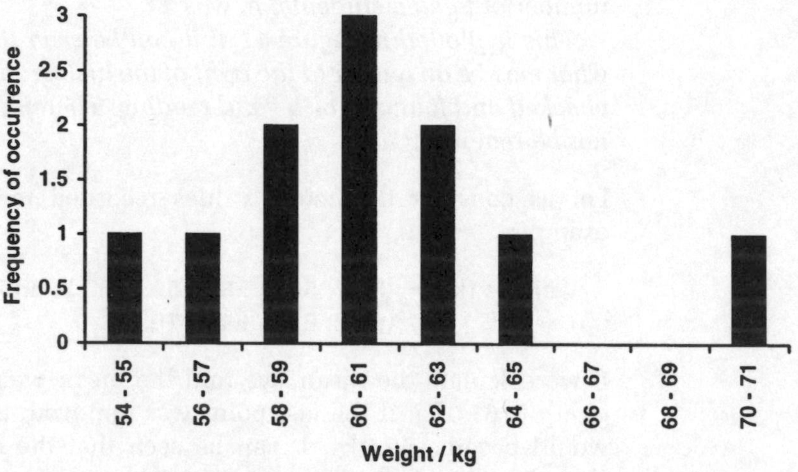

The frequency table shows all 12 samples and the sum of all the values is obtained by summing the product of the frequencies and the quantities:

$$\text{Sum} = (20 \times 1) + (21 \times 1) + (22 \times 4) + (23 \times 3)$$
$$+ (24 \times 2) + (25 \times 1) = 271\,\mu\text{g}(100\,\text{g})^{-1}$$

This can be represented mathematically as in Box 11.3, where f_i represents the frequency of occurrence for the data value x_i. N is the number of classes or sets into which the data have been placed.

Box 11.3 **Sum of a data set recorded in a frequency table**

$$\sum_{i=1}^{N} f_i \times x_i$$

Table 11.4

Vitamin D present (μg/100 g herring)	Frequency
20	1
21	1
22	4
23	3
24	2
25	1

Figure 11.5

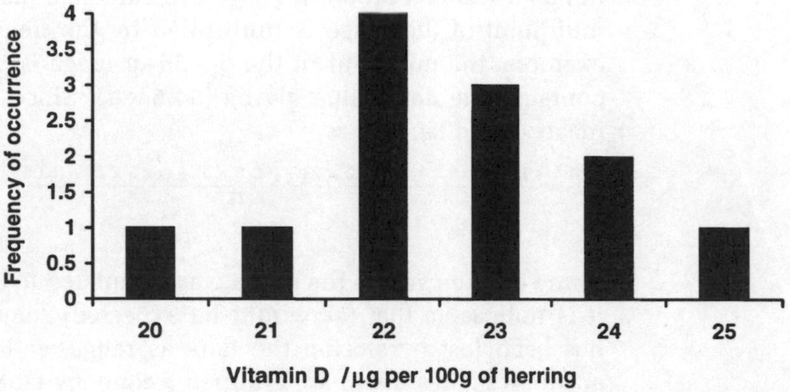

To find the mean, we have to divide the sum by the number of readings, which is simply the sum of the frequency column in the table, i.e. $\sum f_i$, which in this case is 12. So to find the mean from the frequency table, the equation in Box 11.4 is applied.

Box 11.4 **Equation of mean if data set is recorded by frequency.**

$$\text{Mean} = \frac{\sum_{i=1}^{N} f_i \times x_i}{\sum_{i=1}^{N} f_i}$$

In Example 11.2 this is given by:

$$\frac{(20 \times 1) + (21 \times 1) + (22 \times 4) + (23 \times 3) + (24 \times 2) + (25 \times 1)}{12}$$

$$= \frac{271}{12} = 22.6\,\mu g\,(100\,ml)^{-1}$$

Notice that the final mean has been quoted to one significant figure more than the original data. This seems to go against the accurate representation of data (Section 5.3) since usually the final result should only be quoted to the accuracy of the least accurate piece of data. Means are an exception to this rule. If a sample contains more than ten data values and these values have a reasonably small dispersion, the mean can be more accurate than a single measurement, therefore leading to an increase in accuracy of one significant figure.

> If the data set contains more than ten data points, the mean can be represented more accurately than the data values

Notice that Table 11.3 is a grouped frequency table and therefore does not contain the actual readings. In that case the data were recorded in ranges; to calculate the mean, the mid-point of the range is multiplied by the frequency. For example, the midpoint of the 54–55 kg range is 54.5 kg and contains one data value giving (54.5×1). Since $n = 12$ the mean would be:

$$\frac{(54.5 \times 1) + (56.5 \times 1) + (58.5 \times 2) + (60.5 \times 3) + (62.5 \times 2) + (64.5 \times 1) + (70.5 \times 1)}{12}$$

$$= \frac{669.5}{12} = 55.6 \, \text{kg}$$

Using the true values the mean was calculated to be 61 kg, so it is noticeable that (as would be expected) some accuracy has been lost by storing the data as ranges rather than as accurate figures. Even so, grouped frequency tables are useful if many data have to be stored.

11.4.2 *The median and quartiles*

The **median** is the central value in a list of ordered data points. The first step to finding the median is to arrange the data points in order of ascending or descending magnitude. If there is an odd number of data points, the middle value is the median. If there is an even number of points, then the middle two data points should be averaged. The median is also termed the **middle quartile**, since it is the midpoint and an equal number of data values are found above and below this central point. The median is obviously unaffected by outliers but at the same time it makes no use of the actual values represented by the data points. The **upper and lower quartiles** are also often quoted. In the same way as the median is calculated for the 50% mark, the lower quartile corresponds to the 25% mark and the upper quartile corresponds to the 75% mark. The **interquartile range** goes from the lower to the upper quartile and so includes 50% of the data values. When the data points are in order, the median and quartiles can be found using the formula given in Box 11.5.

> The median divides the data set with an equal number of data points above and below it

> The median is unaffected by outliers but makes no use of the actual data values

Box 11.5

$$\text{Median at } \frac{(n+1)}{2}$$

$$\text{Lower quartile at } \frac{(n+1)}{4}$$

$$\text{Upper quartile at } \frac{(3n+1)}{4}$$

For data following a normal distribution, the median will occur near the middle of the curve close to the mean.

The data points can simply be ordered in a line but it may sometimes be useful to arrange them on a **stem and leaf diagram**. This form of diagram is mainly of use if the data points have only two significant figures that vary. The idea is simply to form a 'stem' composed of the first of the variable digits, and then the 'leaves' project out from the stem. This is illustrated in Example 11.3.

Example 11.3

A study was performed to look at haemoglobin levels in the blood of pre-menopausal women. Ten readings were taken and are given in Table 11.5. Find the median and interquartile ranges.

Table 11.5

Patient number	Haemoglobin level/ g(100 ml)$^{-1}$	Patient number	Haemoglobin level/ g(100 ml)$^{-1}$
1	11.1	6	10.6
2	12.3	7	12.4
3	12.1	8	11.9
4	11.8	9	14.1
5	14.2	10	13.5

Forming a stem and leaf diagram:

Stem (first part)	Leaf (second variable digit)
10	6
11	1, 8, 9
12	1, 3, 4
13	5
14	1, 2

The stem and leaf diagram gives a quick way of ordering the data points; furthermore, the leaf section of the diagram acts like a bar chart in giving a visual indication of the distribution. Using the equation from Box 11.5,

The median lies at $(10 + 1)/2 = 5.5$

so between data points five and six,

$$(12.1 + 12.3)/2 = 12.2\,\mathrm{g}(100\,\mathrm{ml})^{-1}$$

Lower quartile is

$$(10 + 1)/4 = 2.75$$

so it lies between data points two and three,

$$(11.1 + 11.8)/2 = 11.45\,\mathrm{g}(100\,\mathrm{ml})^{-1}$$

Since our data are only to one decimal place and only two data points are being considered, this should be represented as $11.5\,\mathrm{g}(100\mathrm{ml})^{-1}$.

Upper quartile is $(3 \times 10 + 1)/4 = 31/4 = 7.75$

so it lies between points seven and eight,

$$(12.4 + 13.5)/2 = 12.95 = 13.0\,\mathrm{g}(100\,\mathrm{ml})^{-1}$$

Using the above quartiles, we know that 50% of our data points lie within the interquartile range, between 11.5 *and* $13.8\,\mathrm{g}(100\mathrm{ml})^{-1}$. *We also know that the middle value is* $12.2\,\mathrm{g}(100\mathrm{ml})^{-1}$.

11.4.3 *The mode*

This is the third commonly used measure of location and distribution. The **mode** corresponds to the most frequently occurring value. If the data are grouped, it is the group with the highest frequency. Sometimes a data set can have more than one mode; for example if there are two values which occur with the same frequency and if these values have the highest frequency of occurrence, then the data set has two modes and is said to be **bimodal**. This term is often used to describe graphs which have two peaks. The mode is not often used in statistical analysis since it depends on the accuracy of the data.

The mode is dependent on the accuracy of the data

11.4.4 *Representing the data with a box plot*

A box–whisker plot is usually used to display large data sets. A rectangular box is drawn, the ends of which represent the upper and lower quartiles. A line is drawn in the box to represent the mean. If the data set follows a normal distribution the data will be symmetrical so the mean will lie between the upper and lower quartiles in the middle of the box. 'Whiskers' are drawn out of the box to record the variability and these show the minimum and maximum values found in the data set. Again, for a normal distribution the whiskers would be of about the same length. An example is given in Figure 11.6.

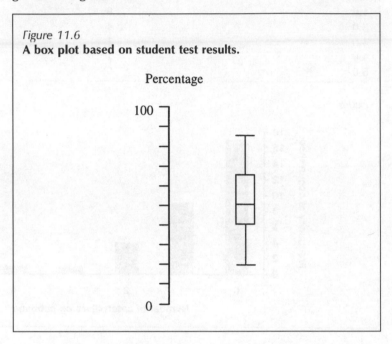

Figure 11.6
A box plot based on student test results.

11.4.5 Mean, median or mode?

As stated in Section 11.4.3, the mode is not widely used since it is dependent on the accuracy of the measurements. Both the mean and the median are used and both give useful information regarding a data set. It is hard to say which of these two measures is the more useful since they give different perspectives on the data set. In general, though, if the data follow a normal or symmetrical distribution, then the mean is a better summary statistic. If the data contain outliers or have a strongly **skewed distribution**, the median may be useful since it is not affected by outliers or skewing. A skewed distribution is one in which the right or left tail is extended, as shown in Example 11.4.

Example 11.4

The number of caterpillars that were infesting a cabbage patch was counted and the data are represented (Table

11.6 and Figure 11.7) as the number of cabbages containing each number of caterpillars from nought to five.

Table 11.6

No. of caterpillars on cabbage	Frequency
0.0	17
1.0	9
2.0	4
3.0	1
4.0	1
5.0	0

Figure 11.7

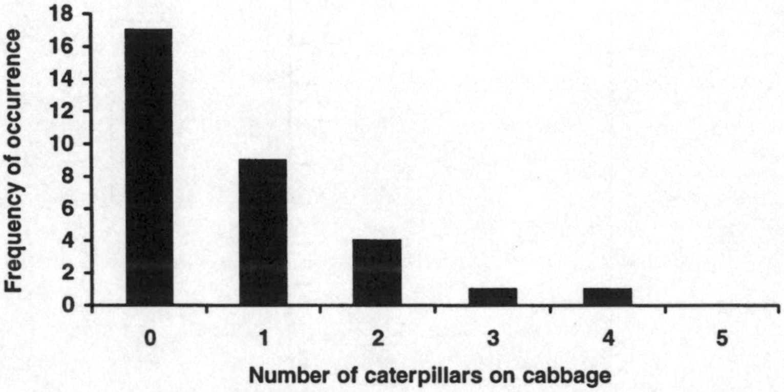

It can be seen that this distribution is not symmetrical and cannot be described as a normal distribution. In this case the mean is greater than the mode, so the graph is said to be **positively skewed**. If the opposite were true and the mode occurred on the far right of the histogram, it would be **negatively skewed**. There are a number of ways of telling whether a distribution is skewed. Probably the best method is to prepare a histogram or a box plot and look at the distribution. A second method is to compare the distance between the mean and the lower quartile with the distance between the mean and the upper quartile. If the distribution is symmetrical these two distances will be the same, but if it is skewed they will not.

Worked examples 11.1

The protein contents were measured in nine common cereals and are listed in g per 100 g of material: {14.0, 10.2, 5.3, 11.0, 7.9, 7.4, 5.3, 9.0, 9.8}.

Find (i) the median, (ii) the mode and (iii) the mean for the data set. Decide whether the data follow a normal or a skewed distribution.

11.5 Measuring variability

The data set has been collected and the mean has been calculated to give a value that is representative of the whole population, but how representative of the population is the mean? Suppose I take the weight of six adult Shih Tzu dogs and at the same time a collegue weighs six different Shih Tzu dogs. This provides two data sets:

Data set 1 (kg) = {4.3, 5.6, 5.6, 5.8, 6.4, 7.1}
Data set 2 (kg) = {5.2, 5.9, 6.0, 6.1, 6.6, 6.8}

$$\bar{x}_1 = 34.8/6 = 5.8\,\text{kg}$$
$$\bar{x}_2 = 36.6/6 = 6.1\,\text{kg}$$

> The accuracy with which the sample mean approximates the population's mean depends on the sample size and variability

So by using the above experiments we have two mean values for the weight of an adult Shih Tzu. If the data sets had been much bigger then the means would have been closer, but in science there are usually limits on how much data can be collected. In the above example we could combine the two data sets to increase our sample size and make the mean more accurate. This gives a mean of 6.0 kg. This highlights the fact that each time you collect data values and take the mean it is likely to be different. The bigger the data sets, the smaller the variation should be. So how confident are we that the mean obtained from our data set is a good estimate of the true mean of the population? The accuracy obviously depends on the sample size and the variability exhibited by the data points. The variability of the data can be found by calculating the variance as described in the following stages.

11.5.1 *Variance*

Once the mean has been calculated, it can be subtracted from each individual value to see how far these values vary from the mean. Since the mean is the central value for a symmetrical distribution, some of these differences will be positive and some negative. Furthermore, if you sum the differences, this will equal zero:

$$\sum_{i=1}^{n}(x_i - \bar{x}) = 0$$

If we take the modulus of each difference (Section 1.3) and then sum them, this will give a measure of variability since the bigger the sum, the greater the variability. This is called the **sum of the differences** (Box 11.6).

Let us consider the Shih Tzu data from Section 11.5.

Box 11.6 **Sum of the differences.**

$$\sum_{i=1}^{n} |x_i - \bar{x}| = \text{sum of the differences}$$

Data set 1:

$$\sum_{i=1}^{n} |x_i - \bar{x}| = |4.3 - 5.8| + |5.6 - 5.8| + |5.6 - 5.8|$$
$$+ |5.8 - 5.8| + |6.4 - 5.8| + |7.1 - 5.8|$$
$$= 1.5 + 0.2 + 0.2 + 0 + 0.6 + 1.3 = 3.8 \text{ kg}$$

Data set 2:

$$\sum_{i=1}^{n} |x_i - \bar{x}| = |5.2 - 6.1| + |5.9 - 6.1| + |6.0 - 6.1|$$
$$+ |6.1 - 6.1| + |6.6 - 6.1| + |6.8 - 6.1|$$
$$= 0.9 + 0.2 + 0.1 + 0 + 0.5 + 0.7 = 2.4 \text{ kg}$$

It can be seen from the above that the first data set has more variability than the second and this agrees with what can be seen by eye. The values in the first set are spread over a greater range. This method can be improved by summing the squares of the differences since this places more weight on outliers that have distorted the mean. At the same time the effects of small differences (less than one) are decreased. This is termed the **sum of the squared differences**.

Box 11.7 **Sum of the squared differences.**

$$\sum_{i=1}^{n} |x_i - \bar{x}|^2 = \text{sum of the squared differences}$$

If we perform sum of squares analysis on the data sets:

Data set 1:

$$\sum_{i=1}^{n} |x_i - \bar{x}|^2 = |4.3 - 5.8|^2 + |5.6 - 5.8|^2 + |5.6 - 5.8|^2$$
$$+ |5.8 - 5.8|^2 + |6.4 - 5.8|^2 + |7.1 - 5.8|^2$$
$$= 2.25 + 0.04 + 0.04 + 0 + 0.36 + 1.69 = 4.38$$
$$= 4.4 \text{ kg}^2$$

Data set 2:

$$\sum_{i=1}^{n} |x_i - \bar{x}|^2 = |5.2 - 6.1|^2 + |5.9 - 6.1|^2 + |6.0 - 6.1|^2$$
$$+ |6.1 - 6.1|^2 + |6.6 - 6.1|^2 + |6.8 - 6.1|^2$$
$$= 0.81 + 0.04 + 0.01 + 0 + 0.25 + 0.49 = 1.6 \, \text{kg}^2$$

It can be seen that by using this method the difference in variability between the two sets is emphasised. The importance of this is perhaps emphasised still further if you consider the two data sets in Example 11.5:

Example 11.5

(a) Mean = 11, values = 9, 13
     Sum of differences = 2 + 2 = 4
     Sum of squares = 4 + 4 = 8

(b) Mean = 4, values = 3, 3, 3, 5
     Sum of differences = 1 + 1 + 1 + 1 = 4
     Sum of squares = 4

The sum of differences was the same for both samples, yet data set (a) contained much greater variability. This was detected by the sum of the squared differences. Example 11.4 also raises another point. As yet, we have not considered the size of the sample. This is taken into account in the calculation of the **variance**.

If you have one data point, you have nothing with which to compare it, so you have no idea what the variability of the population is. If you have two data readings, then you have one estimate of variability, the difference between the two values. With three data points you have two estimates of variability:

(Result2 − Result1) and (Result3 − Result1)

Notice that you do not include the value for (Result3 − Result2) since you already have an idea of how these vary because you know how far each is from Result1. To generalise, if you have n data points you have $n - 1$ *independent* estimates of variability. Here $n - 1$ is termed the **degree of freedom** and is often seen quoted in statistical tests. The variance is therefore measured by the formula in Box 11.8.

Box 11.8 **Equation of variance.**

$$\text{Variance} = \frac{\sum_{i=1}^{n} |x_i - \bar{x}|^2}{n - 1}$$

Variance can be shown to be what statisticians call **unbiased,**
which means it is close to the real variance of the population.
The bigger the value, the greater the variation; but notice that
the units of the variance are the units of the data values
squared.

> The variance gives an
> estimate of the
> variability within the
> population

So for the Shih Tzu data we have:

Data set 1 has a sum of the squared differences of $4.4 \, \text{kg}^2$,
so from Box 11.8:

$$\text{Variance} = 4.4 \div (6 - 1)$$
$$= 4.4 \div 5 = 0.88 \, \text{kg}^2$$

Data set 2 has a sum of the squared differences of $1.6 \, \text{kg}^2$, so
from Box 11.8:

$$\text{Variance} = 1.6 \div (6 - 1)$$
$$= 1.6 \div 5 = 0.32 \, \text{kg}^2$$

If we return to Example 11.5, the effect of dividing by $n - 1$ is
emphasised because of the difference in n:

(a)   Mean $= 11$, values $= 9, 13$
      Sum of differences $= 2 + 2 = 4$
      Sum of squares $= 4 + 4 = 8$
      Variance $= 8 \div (2 - 1) = 8 \div 1 = 8$

(b)   Mean $= 4$, values $= 3, 3, 3, 5$
      Sum of differences $= 1 + 1 + 1 + 1 = 4$
      Sum of squares $= 4$
      Variance $= 4 \div (4 - 1) = 4 \div 3 = 1.3$

11.5.2 *Standard deviation*

Variance gives a good measure of variability but in science
we often want to relate this variability to our mean or data
values. The units of the variance are squared because the

> Variance cannot be
> compared directly with
> the data set because of
> differences in the units

equation contains the sum of the squared differences.
Because the units are squared, the variance cannot be compared directly with the original data. To overcome this problem the square root can be taken (Box 11.9). This is termed
the standard deviation, and if taken from your data can be
represented by the symbol s. If you are referring to the true

deviation, i.e. that seen for the population as a whole, it tends to be donoted by σ.

Box 11.9 **Equation of standard deviation.**

The standard deviation measures the variability in the data

$$\text{Standard deviation} = (\text{variance})^{1/2}$$

$$= \sqrt{\frac{\sum\limits_{i-1}^{n} |x_i - \bar{x}|^2}{n - 1}}$$

You will often see means quoted, plus or minus the standard deviation. This is of relevance because the standard deviation can be related to the normal distribution. Statisticians can show that if the population has a normal distribution, then 68% of the population will occur within one standard deviation of the mean. Within two deviations of the mean you will find approximately 95% of the population and within three deviations 99% of the population. This is shown in Figure 11.8.

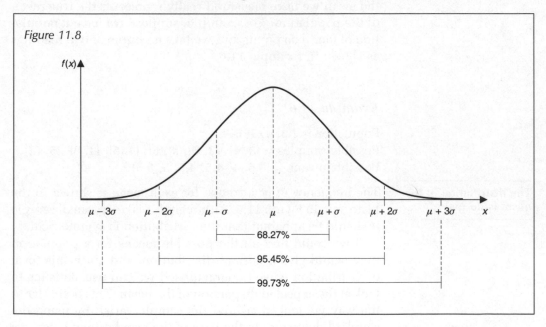

Figure 11.8

Returning to our Shih Tzu data:

Variance of data set 1 = 0.86 kg² so the standard deviation
$$= \sqrt{0.86} = 0.94\,\text{kg}$$

Variance of data set 2 = 0.32 kg² so the standard deviation
$$= \sqrt{0.32} = 0.57\,\text{kg}$$

The mean for data set 2 is 6.1 kg; hence from these data, assuming a normal distribution, which is reasonable since weight is a continuous measure, we expect that 68% of all adult Shih Tzus will have weight 6.1 ± 0.6 kg and 99% of the Shih Tzu population would have a weight of 6.1 ± 1.7 kg. To calculate three times the standard deviation, we have gone back to the more accurate form of the standard deviation (0.57 kg) and rounded to one decimal place after multiplication (Section 5.4).

11.6 Sampling distribution of the mean

The variance and standard deviation give a measure of variability for a data set that has a normal distribution; for example, we know that within one deviation of the mean we should find 68% of the population. But we saw in Section 11.5 that each time we sample the population we are likely to get a different mean. The bigger the sample size, the smaller the variation between means—but how confident are we that the mean we have measured really represents the true mean of the population? For example, suppose we have a population of four data points and we take a sample of two readings as shown in Example 11.6.

Example 11.6

Population = { 3, 4, 5, 6}
Possible samples = {3, 4}, {3, 5}, {3, 6}, {4, 5}, {4, 6}, {5, 6}
Possible means = {3.5, 4, 4.5, 4.5, 5, 5.5}

| The distribution of the mean is normal |

The frequency of occurrence for each mean is shown in the histogram in Figure 11.9. Although this is a very small sample it should be apparent that the distribution is symmetrical.

If we could find all the possible means for a population they would give a normal distribution, and since this form of distribution is well characterised we can use statistics to look at the expected dispersion of the mean. This is similar to the way we looked at how the sample varied by using the standard deviation. In the case of the standard deviation we found that within two deviations of the mean we should find 95% of all the data points. What we now want is to find the mean and say how confident we are, given the data set we have analysed, that the true mean of the population lies between two values. To do this we need to find the **standard error of the mean**.

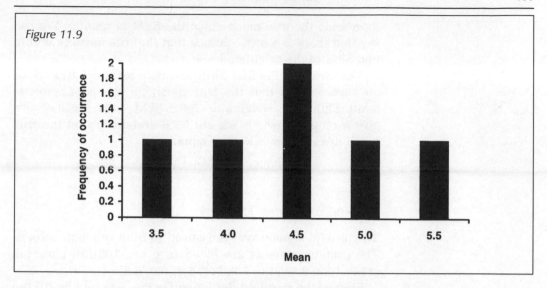

Figure 11.9

11.6.1 *Standard error of the mean*

The standard error of the mean gives an estimate of the expected variability of the sample mean

The standard error of the mean (SEM), or **standard error**, describes the uncertainty about the true value of the population's mean, given that the calculated mean will vary between samples. It is simply obtained by dividing the standard deviation by the square root of the sample size as shown below in Box 11.10.

Box 11.10   **The equation for the standard error.**

$$SEM = (variance/n)^{1/2} = standard\ deviation \div \sqrt{n}$$

$$= \sqrt{\frac{\sum_{i=1}^{n} |x_i - \bar{x}|^2}{n(n-1)}}$$

The SEM therefore decreases as the sample gets bigger, i.e. as the uncertainty decreases.

So for data set 1 of the Shih Tzu weights, we have standard deviation = 0.94 kg, so

$$SEM = 0.94 \div \sqrt{6} = 0.38\ kg$$

For data set 2 of the Shih Tzu weights we have standard deviation = 0.57 kg, so

$$SEM = 0.57 \div \sqrt{6} = 0.23\ kg$$

It can be seen that in the second data set, where we had less variability, we are more confident that the calculated mean represents the true mean since the SEM is smaller. We can say that there is a 68% chance that the true mean is within one SEM of the calculated mean.

The mean for the first Shih Tzu data set was 5.8 kg so we are 68% certain that the true mean for the population of adult Shih Tzu weights is $5.8 \pm \text{SEM} = 5.8 \pm 0.38$ kg. To give a range in which we are 95% certain to find the true mean, we use the following equation:

$$\text{Range} = \text{mean} \pm (\text{SEM} \times 1.96)$$

Worked examples 11.2

The protein content was measured in nine common cereals. The protein contents are listed in g per 100 g of material: {14.0, 10.2, 5.3, 11.0, 7.9, 7.4, 5.3, 9.0, 9.8}.

Find (i) the standard deviation for the data set and (ii) the standard error of the mean.

11.7 Confidence levels and the *t*-distribution

Whenever a mean is calculated there should be an estimate of variability with it, since to appreciate the mean fully we need to know how confident we can be that the population's true mean lies close to this value. If the SEM is given, we can estimate a **confidence interval** for the mean. We saw in Section 11.6 that there is a 68% chance of finding the true mean within one standard error of the mean. This range is therefore called the 68% confidence interval. It is usual to try to be a little more certain than this, so the 95% confidence interval is usually calculated, i.e.:

$$[\text{mean} - (\text{SEM} \times 1.96)] \text{ to } [\text{mean} + (\text{SEM} \times 1.96)]$$

This is a reasonable estimate, but throughout this chapter so far we have assumed that the data set follows a normal distribution. Even if the population has a normal distribution, to ensure that the data representing this population have the same distribution you need a large number of values, at least 30 and preferably more. If the data set contains only a few values, as in the case of Shih Tzu weights where we only had six data points, then although the population as a whole is normally distributed it is likely that the data set does not have a normal distribution. For example, look again at Figure 11.5. Although this was treated as a normal distribution, it looks as though the data are skewed.

If you have less than 30 data values it is usual to use a *t*-**distribution**. This is designed so that as the number of data

For samples with less than 30 data points the *t*-distribution should be used values decreases 't' increases to take account of the increasing uncertainty in your calculated mean. If you multiply the SEM by the appropriate 't' value for the sample size, you can find the 95% confidence range as above. These values are listed in Table 11.7. Notice that the 't' value chosen depends on the degree of freedom for your sample, which corresponds to $n - 1$.

Let us consider the first set of data for the Shih Tzu weights. We calculated that the SEM was 0.38 kg and the

Table 11.7

Degrees of freedom ($n - 1$)	95% probability multiplier	Degree of freedom ($n - 1$)	95% probability multiplier
1	12.706	16	2.120
2	4.303	17	2.110
3	3.182	18	2.101
4	2.776	19	2.093
5	2.571	20	2.086
6	2.447	21	2.079
7	2.365	22	2.074
8	2.306	23	2.069
9	2.262	24	2.064
10	2.228	25	2.059
11	2.201	26	2.056
12	2.179	27	2.052
13	2.160	28	2.048
14	2.145	29	2.045
15	2.131	30	2.042

mean was 5.8 kg. Since we have a small sample size we will work out the 95% confidence interval using the t-distribution. The sample size was six, so the sample has five degrees of freedom. From the above table the 't' value we require to calculate the 95% interval is therefore 2.571.

$$95\% \text{ confidence interval} = 5.8 \pm (0.38 \times 2.571)$$
$$= 5.8 \pm 0.98 \text{ kg}$$

We are therefore 95% certain, based on this data set, that the true mean of the population lies between 4.82 kg and 6.78 kg.

Worked examples 11.3

The protein content was measured in nine common cereals. The protein contents are listed in g per 100 g of material: {14.0, 10.2, 5.3, 11.0, 7.9, 7.4, 5.3, 9.0, 9.8}.
Find the 95% confidence interval for the mean.

Summary

Statistics involves trying to derive parameters which describe a population from a limited set of data points. These data points are assumed to be represenatative of the population. There are three main averages used to describe a population: the mean, the median and the mode.

The mode represents the most frequently occurring data point and is not widely used since it depends on the accuracy of the data. The median is the middle value in the data set when the data points are arranged in order. If the data set contains an even number of values the median is obtained by averaging the two centre values. There are 50% of the data points on either side of the median; the lower quartile is the 25% mark and the upper quartile the 75% mark (Box 11.11). The interquartile range runs between the upper and lower quartiles and contains 50% of all the data points. The median is not affected by outliers but does not use the numerical value represented by the data points.

The mean is a good statistical summary for symmetrical data distributions and is obtained by dividing the sum of the data points by the number of data points.

The mean and median can give different perspectives on a data set and both can be useful. If the data set contains more than ten points and is not too variable, the mean can be quoted to one significant figure more than the data values.

If the data follow a symmetrical bell-shaped curve the median, mode and mean all occur in the middle, but if it is skewed the values will be separated. The skewing can be seen by comparing the difference between the mean and the upper quartile with the difference between the mean and the lower quartile. For a symmetrical distribution these will be the same.

The variability of the data set can be estimated by the variance. This is thought to give a good approximation to the variability of the population as it is unbiased. To relate the variance to the data, the square root must be taken and this provides the standard deviation: 68% of all the data points occur within plus or minus one deviation from the mean, 95% occur within two deviations and 99% within three (Box 11.12).

Box 11.12

$$\text{Variance} = \frac{\sum\limits_{i=1}^{n} |x_i - \bar{x}|^2}{n-1}$$

$$\text{Standard deviation} = \sqrt{\text{variance}}$$

$$= \sqrt{\frac{\sum\limits_{i=1}^{n} |x_i - \bar{x}|^2}{n-1}}$$

Each time a population is sampled, a different mean may be obtained. The means for a population follow a normal distribution; therefore the potential variability in the mean, given the data set from which it was derived, can be calculated by finding the standard error of the mean, SEM (Box 11.13). It is 68% certain that the population's true mean will lie within plus or minus one SEM from the mean. 95% confidence can be obtained using the following equation:

Box 11.11

$$\text{Mean at } \sum_{i=1}^{n} \frac{x_i}{n}$$

$$\text{Median at } \frac{(n+1)}{2}$$

$$\text{Lower quartile at } \frac{(n+1)}{4}$$

$$\text{Upper quartile at } \frac{(3n+1)}{4}$$

Box 11.13

$$SEM = (variance \div n)^{1/2}$$
$$= standard\ deviation \div \sqrt{n}$$

95% confidence interval
= mean ± (SEM × 1.96)

If there are less than 30 data points, then although the population being modelled may have a normal distribution, the data themselves are unlikely to have a normal distribution because there are too few values. In this case the t-distribution can be used. To find the 95% confidence interval, the SEM is multiplied by the correct 't' value from Table 11.7 This value depends on the number of data points in the sample.

End of unit questions

1. The birth weight of 15 babies was recorded and the data set is shown below in kg:

    {3.9, 3.7, 4.0, 3.2, 3.7, 2.9, 4.4, 2.7, 3.0, 4.2, 4.2, 2.6, 3.8, 3.3, 3.7}

    (a) Find the mean, median and mode.

    (b) In what weight range would you expect to find 95% of newborn babies?

2. The forced expiratory volume (FEV$_1$) is a diagnostic measure used in respiratory medicine to determine if a patient is asthmatic. The FEV$_1$ will vary with age so the result is displayed as a percentage of the value you would expect to obtain from a healthy individual. The following values were obtained from men suffering from pneumoconiosis:

    {48, 70, 83, 54, 62, 94, 67, 74, 86, 102}

    (a) Produce a box plot and decide if the data follow a normal distribution.

    (b) What is the mean for the above data? Give the 95% confidence interval for your answer.

3. Monolayer tanks can be used to mimic a membrane environment. A peptide or drug is placed in the tank below a single layer of lipid. If the drug or peptide inserts into the lipid it causes the pressure to increase and this can be detected. Two peptides were tested to see if they could insert into the lipid. The results are given in Table 11.8.

    (a) Calculate the mean pressure change for each peptide.

    (b) Give the 95% confidence interval for the means.

Table 11.8

Peptide	Pressure change/mN m^{-2}					
1	11.08	12.01	11.5	12.7		
2	2.4	2.8	1.9	2.6	2.2	2.4

Source: Adapted from M. J. Campbell and D. Machin (1993), *Medical Statistics*, 2nd edn. New York: John Wiley.

Appendix Solutions to Problems

Worked examples

Chapter 1

Examples 1.1

   (i) $2\times -5 = -10$

  (ii) $-6\times-3 = 18$

 (iii) $3 - 5 = -2$

 (iv) $-2 - 6 = -8$

  (v) $-3 - (-4) = -3 + 4 = 1$

 (vi) $-6 \div -6 = 1$

(vii) $6 \div -12 = -0.5$

Examples 1.2

   (i) $-2 - |-2| = -2 - 2 = -4$

  (ii) $|3 - 5| = |-2| = 2$

 (iii) $1 - 4 - |3| = 1 - 4 - 3 = -6$

 (iv) $3 + |2 - 3| = 3 + |-1| = 3 + 1 = 4$

Examples 1.3

   (i) $3 - 9 \div 3 = 3 - 3 = 0$

  (ii) $4 \times (2 - 3) = 4 \times -1 = -4$

 (iii) $((4 + 6) \div 5 + 3) \times 3 = (10 \div 5 + 3) \times 3$
        $= (2 + 3) \times 3 = 5 \times 3 = 15$

 (iv) $10 \times 5 + 4 \times 5 = 50 + 20 = 70$

  (v) $((15 - 5) + 2 \times 2) \div 7 = (10 + 2 \times 2) \div 7$
       $= (10 + 4) \div 7 = 14 \div 7 = 2$

Examples 1.4

   (i) $18 \times 32 \div 9 = 18 \div 9 \times 32 = 2 \times 32 = 64$

  (ii) $55 \div 13 \times 26 = 55 \times 26 \div 13 = 55 \times 2 = 110$

 (iii) $(16 + 17) \div 11 \div 6 = 33 \div 11 \div 6 = 3 \div 6 = 0.5$

Chapter 2

Examples 2.1

(i) $9/36 = \frac{1}{4}$

(ii) $27/18 = 1\frac{9}{18} = 1\frac{1}{2}$

(iii) $24/16 = 1\frac{8}{16} = 1\frac{1}{2}$

(iv) $\frac{3}{7}$

Examples 2.2

(i) 168 (ii) 198 (iii) 54 (iv) 792

Examples 2.3

(i) $1/3 + 7/8 = 8/24 + 21/24 = 29/24$ or $1\frac{5}{24}$

(ii) $1/2 - 4/10 = 5/10 - 4/10 = \frac{1}{10}$

(iii) $5/7 - 10/12 = 60/84 - 70/84 = -10/84 = -\frac{5}{42}$

(iv) $3/4 \times 2/7 = \frac{6}{28} = \frac{3}{14}$

(v) $4/11 \times 22/30 = 4/1 \times 2/30 = 2/1 \times 2/15 = \frac{4}{15}$

(vi) $6/13 \div 1/2 = 6/13 \times 2/1 = \frac{12}{13}$

(vii) $2/3 \div 1/9 = 2/3 \times 9/1 = 2/1 \times 3/1 = 6/1 = 6$

Examples 2.4

(a) $1 - \frac{7}{10} = \frac{3}{10}$. Therefore $3/10 \times 100 = 30\%$ remains

(b) (i) 75% (ii) 66.6% (iii) 50% (iv) 52.94% (v) 92.86%

(c) (i) 16 (ii) 7.7 (iii) 13.28 (iv) 11.16

Examples 2.5

(a) (i) A = 20 ml, B = 40 ml, C = 40 ml

(ii) A = 50 ml, B = 50 ml

(iii) A = 10 ml, B = 40 ml, C = 30 ml, D = 20 ml

(iv) B = 33.3 ml, C = 16.7 ml, D = 50 ml

(b) (i) A : B : C in the ratio 6 : 1 : 5

(ii) A : B in the ratio 1 : 3

(iii) A : B : C in the ratio 13 : 6 : 3

(iv) A : B : C in the ratio 5 : 2 : 4

(c) (i) 4 ml (ii) 2 ml

Chapter 3

Examples 3.1

(i) $t - (2t + c) = t - 2t - c = -t - c$

(ii) $p + c - p = c$

(iii) $xy + 2x - y + 4xy = 5xy + 2x - y$

(iv) $z + (t - c) = z + t - c$

(v) $-2(3 - y) = -6 + 2y = 2y - 6$

Examples 3.2

(i) Common factors are 2 and 3 so highest common factor
$= 2 \times 3 = 6$

(ii) There are no common factors

(iii) Common factors are 2 and 11 so highest common factor
$= 2 \times 11 = 22$

(iv) Common factors are 2, 3 and 3 so highest common
factor $= 2 \times 3 \times 3 = 18$

(iv) Common factors are 3, 3 and 3 so highest common
factor $= 3 \times 3 \times 3 = 27$

Examples 3.3

(i) $2ab \div (ab + 3ab) = 2ab \div ab(1 + 3) = 2 \div (1 + 3)$
$= 2/4$ or $\frac{1}{2}$

(ii) $3x \div (6 - 18x) = 3x \div 3(2 - 6x) = x \div (2 - 6x)$ or
$x/(2 - 6x)$

(iii) $ab \div (ab + a) = ab \div a(b + 1) = b \div (1 + b)$ or b/(1+b)

(iv) $3a/6b \times 3b/a = 3/6b \times 3b/1 = 3/2b \times b/1$
$= 3/2 \times 1/1 = 3/2$ or $1\frac{1}{2}$

(v) $3/2 \times t/7 = 3t/14$

(vi) $\dfrac{2}{(a + b)} - \dfrac{6}{b} = \dfrac{2b}{b(a + b)} - \dfrac{6(a + b)}{b(a + b)} = \dfrac{2b - 6a - 6b}{b(a + b)}$

$= \dfrac{-6a - 4b}{b(a + b)} = \dfrac{-2(3a + 2b)}{b(a + b)}$

(vii) $\dfrac{3xy}{t} + \dfrac{7xy}{2m} = \dfrac{6mxy}{2tm} + \dfrac{7txy}{2tm} = \dfrac{xy(6m + 7t)}{2tm}$

Examples 3.4

(i) $y = 2/x$ so $x = 2/y$

(ii) $y = 7/(x - 3)$ so $y(x - 3) = 7$ so $x - 3 = 7/y$ so
$x = (7/y) + 3$

(iii) $y = (x - 6) - 2$ so $y + 2 = x - 6$ so $x = y + 2 + 6$ so
$x = y + 8$

(iv) $2 = 3xy$ so $y = 2/(3x)$

Examples 3.5

(a) (i) $-1 \leq x < 3$ (ii) $6 < x < 11$ (iii) $0 < x \leq 8$
    (iv) $4 \leq x \leq 5$

(b) (i) $x - 3 > 2$ so $x > 5$

   (ii) $6 - x > 4$ so $-x > -2$ so $x < 2$

   (iii) $7 + x \geq 6$ so $x \geq -1$

Examples 3.6

(i) $k_1[A][B] = k_{-1}[P][Q]$; $v_f = k_1[A][B]$ and $v_f = v_r$ so
$v_r = [P][Q]$

(ii) $v_f = k_1[A][B]$ so $k_1 = v_f/([A][B])$
$= \text{mol litre}^{-1} \text{min}^{-1}/(\text{mol litre}^{-1} \times \text{mol litre}^{-1})$
$= \text{min}^{-1}/(\text{mol litre}^{-1}) = \text{mol}^{-1} \text{litre min}^{-1}$

Chapter 4

Examples 4.1

(i) $5^3 = 5 \times 5 \times 5 = 125$

(ii) $2^{-5} = 1/(2 \times 2 \times 2 \times 2 \times 2) = 1/32$

(iii) $(-5)^2 = 25$

(iv) $(-2)^5 = -(2^5) = -32$

(v) $1.1147^9 = 2.6572$

(vi) $(-5.73)^5 = -(5.73)^5 = -6176.9$

Examples 4.2

(a) (i) Add indices to give $2^{2+2} = 2^4 = 16$

    (ii) Add indices to give $2^{3+(-3)} = 2^0 = 1$

    (iii) Combine indices to give

$$2^{2+4-(-3)+(-4)} = 2^{2+4+3-4} = 2^5 = 32$$

    (iv) Subtract indices to give $3^{2-5} = 3^{-3} = 1/27$

    (vi) Subtract indices to give $106^{11-8} = 106^3 = 1191016$

(b) (i) 6^{11} (ii) z^3 (iii) $c^0 = 1$ (iv) $a^{-5} = 1/a^5$ (v) c^7

Examples 4.3

(a) (i) $2^4 = 16$ (ii) $2^{-9} = 1/2^9 = 1/512$ (iii) $4^{10} = 1048576$

(b) (i) a^{36} (ii) e^{-8} (iii) e^6

Examples 4.4

(a) (i) 2.39×10^2 (ii) 3.6×10^{-3}

   (iii) $(2 \times 10^2) \times (3 \times 10^{-6}) = 6 \times 10^{-4}$

   (iv) 9.73 or 9.73×10^0

   (v) $(1.792 \times 10^3) \times (1.792 \times 10^{-4}) = 1.792^2 \times 10^{-1}$
   $= 3.211 \times 10^{-1} = 0.3211$

(b) (i) 10×10^{-6} or 1×10^{-5} litres

   (ii) 10×10^{-3} or 1×10^{-2} ml

(c) (i) 1×10^4

   (ii) $(10 \times 10^{-6}) \times (1 \times 10^4) = 10 \times 10^{-2}$
   $= 1 \times 10^{-1}$ litre $= 0.1$ litre or 10^2 ml

   (iii) Four

Chapter 5

Examples 5.1

(a) (i) 10% (w/v) (ii) 6.67% (w/v) (iii) 15% (w/v)

   (iv) 10% (w/v)

(b) (i) $5(5 + 48) = 9.4\%$ (w/w)

Examples 5.2

(i) 20 mM $= 20 \times 10^{-3}$ mol litre^{-1}
1 mol $= 58.5$ g so 20 mmol $= 58.5 \times (20 \times 10^{-3})$
$= 1.17$ g so 20 mM $= 1.17$ g litre^{-1}

(ii) 1 litre of 20 mM requires 1.17 g, 5 ml of 20 mM requires
$(5/1000) \times 1.17 = 5.85$ mg

(iii) 1 mol $= 58.5$ g; 1 μmol $= 58.5$ μg

Examples 5.3

(a) (i) 23.3 (ii) $129\,000$ (iii) $0.003\,43$ (iv) $267\,000$

(b) (i) 45.096 (ii) 0.465 (iii) 0.001 (iv) 1289.632

Examples 5.4

(i) $12.354 \times 3.23 = 39.90342 = 39.9$ (3 significant figures)

(ii) $5 + 4.35 \times 2.3 = 15.005 = 20$ (1 significant figure)

(iii) $3.00 \times 2.34 \div 4.001 = 1.75456136 = 1.75$ (3 significant figures)

Chapter 7

Examples 7.1

(a) $f(4) = 2 \times 4^2 - 5 = 27$
$f(0) = 2 \times 0^2 - 5 = -5$
$f(-3) = 2 \times (-3)^2 - 5 = 13$
$f(5)$ lies outside the domain so is not defined for this function.

(b)   (i) $f(x): (3x - 2) \div 6$

(ii) $f(x): 8 - 5x^2$

Examples 7.2

(i) $f(x) = 7x + 3$ so $y = 7x + 3$
$x = (y - 3) \div 7$
$f^{-1}(x) = (x - 3) \div 7$

(ii) $f(x) = 3 - x$ so $y = 3 - x$
$x = 3 - y$
$f^{-1}(x) = 3 - x$

(iii) $f(x) = 1/x$ so $y = 1/x$
$x = 1/y$
$f^{-1}(x) = 1/x$

(iv) $t(x) = -5/(6x)$ so $y = -5/(6x)$
$x = -5/(6y)$
$t^{-1}(x) = -5/(6x)$

Examples 7.3

(i) Gradient $= m = (5 - 2) \div (2 - 0) = 3/2 = 1.5$
Substituting into the equation of a straight line for the ordered pair $(0, 2)$,
$2 = (1.5 \times 0) + c$ so $c = 2$.
The equation is therefore $y = 1.5x + 2$

(ii) Substituting into the equation of a straight line:
$4 = 4m + 3$ so $m = (4 - 3) \div 4 = 0.25$
The equation is therefore $y = 0.25x + 3$

Examples 7.3

(i) $t - 3 = 0$ so $t = 3$

(ii) $5a = 2a + 3$ so $3a = 3$ and therefore $a = 1$

(iii) $x + 9 = 2$ so $x = -7$

(iv) $2 = 1/(x - 7)$ so $2(x - 7) = 1$ so $x = 1/2 + 7 = 7.5$

Examples 7.5

(i) Absorption is proportional to concentration; therefore three-fold dilution would decrease absorption three-fold to 0.2 units

(ii) Absorption is proportional to path length, so doubling the path length doubles the absorption to 1.2 units

(iii) The increase in path length and decrease in concentration cancel each other out so the absorption remains at 0.6 units

Examples 7.6

(a) $$1/v = (K_m/V_{max}) \times 1/[S] + 1/V_{max}$$
$$[S]/v = ([S]K_m/V_{max}[S]) + [S]/V_{max}$$
$$= (K_m/V_{max}) + [S]/V_{max}$$
so $[S]/v = (1/V_{max})[S] + K_m/V_{max}$

(b) $y = [S]/v$ and $x = [S]$ so these values would be plotted.

(c) The gradient $m = 1/V_{max}$ and so can give V_{max} since $V_{max} = 1/m$. The y intercept is K_m/V_{max}. V_{max} is known $(V_{max} = 1/m)$ so K_m can be found.

Chapter 8

Examples 8.1

(i) $x^2 + 5x - 6 = (x - 1)(x + 6)$ so $x = 1$ or -6

(ii) $-2x^2 - x + 3 = (2x + 3)(-x + 1)$ so $x = 1$ or -1.5

(iii) $x(1 - x) = x(2x - 1)$ so $x - x^2 - 2x^2 + x = 0$
$-3x^2 + 2x = 0$ so $x(-3x + 2) = 0$ so $x = 0$ or $x = 2/3$

Examples 8.2

(i) Using the formula with $a = 2$, $b = -6$, $c = 4$ we find the discriminant $= 4$ so there are two roots, $x = 2$ or 1

(ii) Using the formula with $a = 1$, $b = 4$, $c = -8$ we find the discriminant $= 48$ so there are two roots, $x = 1.46$ or -5.46

(iii) Using the formula with $a = 2$, $b = -7$, $c = 3$ we find the discriminant $= 25$ so there are two roots, $x = 3$ or 0.5

Chapter 9

Examples 9.1

(a)   (i) $4x + x/2 = (9x)/2$ and $4x + (-x) = 3x$ so
$$e^{4x}(e^{x/2} + e^{-x}) = e^{9x/2} + e^{3x}$$

    (ii) $8x \times 1/2 = 4x$ so $(e^{8x})^{1/2} = e^{4x}$

    (iii) $x - 2x = -x$ and $7x - 2x = 5x$ so
$$(e^x + e^{7x})/e^{2x} = e^{-x} + e^{5x}$$

    (iv) $e^x - (e^{3x})^2 = e^x - e^{6x}$

(b)   (i)   3.32   (ii) 0.50   (iii) 1.22   (iv) $1/e^3 = e^{-3} = 0.05$

Examples 9.2

(i) A graph of $y = e^x$ and $y = 2x + 3$ intercept at approximately $x = 1.9$ as shown in Figure A.1, so this is the solution.

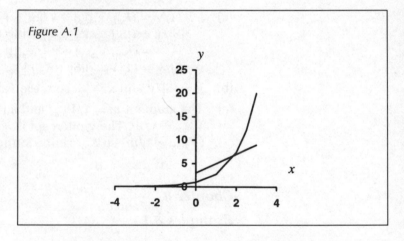

Figure A.1

(ii) The graph of $y = e^x$ and $y = 6.7$ intercept at approximately $x = 1.9$ so this is the solution.

(iii) The graphs of $y = e^x$ and $y = e^{-x} + 10$ intercept at approximately $x = 2.32$ so this is the solution.

Chapter 10

Examples 10.1

  (i) $\log_{10} 100 = \log_{10} 10^2 = 2$

  (ii) $\log_{10} 10^7 = 7$

  (iii) $\log_{10} 1 = \log_{10} 10^0 = 0$

  (iv) $\log_{10} 10^{2.3} = 2.3$

Examples 10.2

(i) $\log_{10} 10 = \log_{10} 10^1 = 1$

(ii) $\log_2 8 = \log_2 2^3 = 3$

(iii) $\log_5 125 = \log_5 5^3 = 3$

(iv) $\log_4 16 = \log_4 4^2 = 2$

Examples 10.3

(a) (i) $\log 2 + \log 6 = \log(2 \times 6) = \log 12$

(ii) $3 \log 2 - 2 \log 4 = \log 2^3 - \log 4^2 = \log(8/16) = \log 0.5$

(iii) $2 \log a - \log 6 = \log(a^2/6)$

(b) (i) $\log 2^6 = 6 \log 2 = 6 \times 0.3 = 1.8$

(ii) $\log 12 = \log(2 \times 6) = \log 2 + \log 6 = 0.3 + 0.78 = 1.08$

(iii) $\log 36 = \log 6^2 = 2 \log 6 = 2 \times 0.78 = 1.56$

(iv) $\log 3 = \log(6/2) = \log 6 - \log 2 = 0.78 - 0.3 = 0.48$

Examples 10.4

(a) $pH = -\log_{10}$ [hydrogen ions concentration]

(i) $pH\,5 = 10^{-5}\,M = 10\,\mu M$

(ii) $pH\,7.4 = 10^{-7.4}\,M = 4.0 \times 10^{-8}\,M = 40\,nM$

(iii) $pH\,10.2 = 10^{-10.2}\,M = 6.3 \times 10^{-11}\,M = 63\,pM$

(iv) $pH\,2.9 = 10^{-2.9}\,M = 1.26 \times 10^{-3}\,M = 1.26\,mM$

(b) Using the equation in question (a):

(i) $pH = 3$ (ii) $pH = 9.96$ (iii) $pH = 4$ (iv) $pH = 7.1$

Chapter 11

Examples 11.1

(i) If the data are arranged in order the median
$= (n+1)/2 = 5$
Data set $= \{5.3, 5.3, 7.4, 7.9, 9.0, 9.8, 10.2, 11, 14\}$
The fifth data point is 9.0 g per 100 g of material so this is the median.

(ii) Mode $= 5.3$ g per 100 g of material

(iii) Mean $= 79.9/9 = 8.9$ g per 100 g of material
The median and mean are very close; hence even though the mode occurs at one end of the data set, this would appear to have a normal rather than a skewed distribution.

Examples 11.2

  (i) Standard deviation = 2.79

 (ii) Standard error of the mean = 0.93

Examples 11.3

Using Table 11.7, for $n - 1 = 8$,

$$95\% \text{ confidence interval} = \text{mean} \pm (\text{standard error of mean}$$
$$\times 1.96)$$
$$= 8.9 \pm (0.93 \times 1.96)$$
$$= 8.9 \pm 1.82 \,\text{g per 100 g}$$
$$\text{of material}$$

End of unit questions

Chapter 1

1. (a) $20 \times 18.5 \times 5 = 20 \times 5 \times 18.5 = 100 \times 18.5 = 1850$

    (b) $0.6 \times 12.5 \times 5 \times 8 = 12.5 \times 8 \times 0.6 \times 5$
$$= 100 \times 0.6 \times 5 = 60 \times 5 = 300$$

    (c) $32 \times 5 \div 8 = 32 \div 8 \times 5 = 4 \times 5 = 20$

2. (a) $4 - 7 = -3$

    (b) $-3 - (-2) = -3 + 2 = -1$

    (c) $9 + 23 - 47 - 2 = -47 - 2 + 9 + 23$
$$= -49 + 9 + 23 = -40 + 23 = -17$$

3. (a) $a \times -b = -ab$

    (b) $a \times -b \times -c = abc$

    (c) $-c \times -b = cb$

4. (a) $(6 - 2) \div 4 + 7 = 4 \div 4 + 7 = 1 + 7 = 8$

    (b) $22 \times 7 \div 11 + 6 - 3 = 22 \div 11 \times 7 + 6 - 3$
$$= 2 \times 7 + 6 - 3 = 20 - 3 = 17$$

    (c) $(((24 - 14) - 5 \times 6) - 5) + 25 - 40 \div 8$
$$= ((10 - 5 \times 6) - 5) + 25 - 40 \div 8$$
$$= ((10 - 30) - 5) + 25 - 40 \div 8$$
$$= (-20 - 5) + 25 - 40 \div 8$$
$$= -25 + 25 - 40 \div 8$$
$$= -25 + 25 - 5$$
$$= -5$$

5.

Estimate	10	16	19	23
Error	−6.3	−0.3	2.7	6.7
Absolute error	6.3	0.3	2.7	6.7

6. (a) 0.000014

   (b) $(0.037 \times 0.037 \times 0.01) \div (1 - 0.037)$

7. (a) 1.7685

   (b) $(20\,000 \times (273 \div 310) + 3000 \times 0.024) \div 10\,000$

Chapter 2

1. (a) $1/2 + 5/7 = 7/14 + 10/14 = 17/14$ or $1\frac{3}{14}$

   (b) $2/6 - 1/4 = 4/12 - 3/12 = \frac{1}{12}$

   (c) $6/7 \times 2/3 = 2/7 \times 2/1 = \frac{4}{7}$

   (d) $1/9 \div 4/3 = 1/9 \times 3/4 = 1/3 \times 1/4 = \frac{1}{12}$

2. (a) $A = 25\%$, $B = 62.5\%$, $C = 12.5\%$

   (b) $A = 12.5\%$, $B = 29.2\%$, $C = 58.3\%$

3. (a) 1.446 m (b) 11.24%

4. (a) Corn = 23.3%, wheat = 7.78%, barley = 38.89%

   (b) Corn = 46.67 acres, wheat =15.56 acres,
       barley = 77.78 acres

5. 40% secreted, so $0.28 \times 60 = 16.8\%$ metabolised, and
   43.2% remains.
   Secreted : metabolised : remaining gives a ratio of 40 :
   16.8 : 43.2.

6. Chloroform = 156.25 ml; methanol = 84.14 ml; water =
   9.62 ml

7. (a) Every 14.3 days the sample decreases by 50% so
       we have the following:

Time in multiples of 14.3 days	Amount of sample left (%)
0	100
1	$100 \times 50/100 = 50$
2	$50 \times 50/100 = 25$
3	$25 \times 50/100 = 12.5$
4	$12.5 \times 50/100 = 6.25$

   Hence the sample decays to 6.25% in
   $14.3 \times 4 = 42.9$ days.

   (b) Using the same principle as in section (a), 99.6%
       has decayed so $11226 \times 0.4/100 = 44.9\,Bq$ remain.

8. (i) $3.09\,\mu m$ or $2.91\,\mu m$ (ii) $3.3\,\mu m$ or $2.7\,\mu m$

   (iii) $3.24\,\mu m$ or $2.76\,\mu m$ (iv) $3.45\,\mu m$ or $2.55\,\mu m$

   (v) $3.03\,\mu m$ or $2.97\,\mu m$

9.

Amino acid	% (w/w)	Amount in 0.5 g (g)
Alanine	4.6	0.023
Arginine	3.1	0.016
Glycine	5.2	0.026
Leucine	13.6	0.068
Valine	19.4	0.097

10.   Myristic acid $= 6.43\,\mu M$; palmitic acid $= 17.14\,\mu M$; oleic acid $= 25.71\,\mu M$

Chapter 3

1.   (a)   The units are min^{-1}.

     (b)   The two rate constants cannot be compared since the units are different.

2. $V_{max} = 0.000\,021\,mol\,litre^{-1}\,min^{-1}$

3.   Substituting into the equation gives 70 253 Da.

4.   (a)   Substituting into the equation gives 1.769 (Note: V_g and V_f are given in cm^3 and the equation requires mm^3, where $1\,cm^3 = 1000\,mm^3$; also $V_g = 23\,cm^3$).

     (b) $\alpha = 0.019$

5.   (a)   (i) $w < 30$ (ii) $30 \le w < 35$ (iii) $35 \le w < 40$
         (iv) $w \ge 40$

     (b)   Group (ii): $[30, 35)$; Group (iii): $[35, 40)$

6.   If x represents the number of visits in one hour:
     no visits, $x = 0$ $[1, 5)$, $1 \le x < 5$
     ten or more visits, $x \ge 10$ $[5, 10)$, $5 \le x < 10$

7. $K_i = 0.000\,006\,5\,mol\,litre^{-1}$

Chapter 4

1.   13 orders of magnitude

2.   (a)   Diameter $= 4 \times 10^{-6}\,m$, therefore
         radius $= 2 \times 10^{-6}\,m$
         Substituting into the equation:
         volume $= 3.35 \times 10^{-17}\,m^3$

     (b)   Volume decreases by $0.35 \times (3.35 \times 10^{-17})$
         $= 1.17 \times 10^{-17}$ so new volume is:
         $3.35 \times 10^{-17} - 1.17 \times 10^{-17} = 2.18 \times 10^{-17}\,m^3$
         Transposing the formula to make r the subject, we
         get $r = 1.73 \times 10^{-6}\,m$ or $1.73\,\mu m$ so the diameter is
         $3.46\,\mu m$.

3. $(9 \times 10^{12}) \div (3.2 \times 10^{-2}) \times (6.02 \times 10^{23}) = 1.69 \times 10^{38}$

4. $l : h^{2/3}$ is 35 : 5.24 or 6.68 : 1 so the ratio holds.

5. $(h^2)^{1/3} = h^{(2 \times 1/3)} = h^{2/3}$

6. $(1 \times 10^{-12}) + (4 \times 10^{-12}) + (2 \times 10^{-10}) = 2.05 \times 10^{-10}$ s.

7. (a) 50 nucleotides s^{-1} so 1.062×10^3 nucleotides take $1.062 \times 10^3 \div 50 = 21.24$ s

   (b) One mistake every 100 bases so (1.062×10^3) $\div (1 \times 10^2) - 1.062 \times 10^1$ per mutant bacterium.

   (c) One mistake in 10^5 bases, therefore (1×10^5) $\div (1.062 \times 10^3) = 94.16$ so the gene would have to be transcribed approximately 94 times to expect one mistake.

8. $(1 \times 10^{-14}) \div (5 \times 10^{-10}) = (1 \div 5) \times (10^{-14} \div 10^{-10})$
   $= 2 \times 10^{-5}$ mol litre^{-1}

Chapter 5

1. (a) Glucose = 0.4% (w/v), potassium dihydrogen-phosphate = 1% (w/v), magnesium sulphate = 0.2% (w/v), citric acid = 0.2% (w/v)

   (b) Glucose = 4/180 = 0.022 M or 22 mM, potassium dihydrogenphosphate = 10/136 = 0.074 M = 74 mM, magnesium sulphate = 0.2/120 M = 1.7 mM, citric acid = 2/192 M = 10.4 mM

2. Tryptone = $1.7 \times (30/100) = 1.7 \times 0.3 = 0.51$ g
   Peptone = $0.3 \times 0.3 = 0.09$ g or 90 mg
   Glucose = $0.25 \times 0.3 = 0.075$ g or 75 mg
   Sodium chloride = $0.5 \times 0.3 = 0.15$ g

3. 3 litres of 0.5 M acid requires $3 \times 0.5 = 1.5$ mol of material
   1 mol = 36.5 g so 1.5 mol = 54.75 g.
   The stock contains 25 g of acid in each 100 g of stock, i.e each gram only contains 0.25 g of acid. We therefore require: $54.75 \div 0.25 = 219$ g of stock to get the required amount of acid.
   Density = mass/volume so 1.15 g ml^{-1} = 219 g/volume
   We therefore require $219 \div 1.15 = 190.4$ ml of acid.
   This can be made up to 3 litres.

4. (a) 23.48 g in 180 ml, so there is 13.04 g in 100 ml = 13.04% (w/v)

   (b) $(23.48 \div 238.31) \times (180 \div 1000) = 0.018$ M or 18 mM

5. $0.4\,M = 0.4 \times 155.2 = 62.1\,\mathrm{g\,litre}^{-1}$ or 6.25 g per 100 ml
    $= 6.21\%$ (w/v), so the stock can be used in diluted form.
    The reaction requires $(0.05/6.21) \times 100 = 0.008\,\mathrm{ml}$ of
    stock solution made up to 100 ml.

6. (a) $1\,M = 40\,\mathrm{g\,litre}^{-1}$ so $0.03\,M = 0.03 \times 40 = 12\,\mathrm{g}$
         $\mathrm{litre}^{-1} = 3.6\,\mathrm{g}$ in 300 ml

    (b) 3.6 g in 300 ml $= 1.2\,\mathrm{g}$ in 100 ml $= 1.2\%$ (w/v)

7. To convert 5 M to 5 mM there needs to be a 1000-fold
    dilution; therefore take 1.5 ml and make up the volume
    to 1500 ml.

8. Density = mass/volume so $0.79\,\mathrm{g\,ml}^{-1} = \mathrm{mass}/100\,\mathrm{ml}$
    so mass $= 0.79 \times 100 = 79\,\mathrm{g}$
    1 mol $= 46\,\mathrm{g}$ so $79\,\mathrm{g} = 1.72\,\mathrm{mol}$

9   (a) $1\,\mathrm{mg\,ml}^{-1} = 1\,\mathrm{g\,litre}^{-1}$ so
         molarity $= 1/75.07 = 0.013\,M = 13\,\mathrm{mM}$

    (b) $(10 \times 10^{-6} \div 0.013) \times 10 = 0.0077\,\mathrm{ml}$ or 7.7 µl
         need to be made up to 10 ml.

Chapter 6

1. Relative preferences of lysine (Fig. A.2) are:
    Helix $= 1.09/3.35 = 0.32$
    Sheet $= 0.42/3.35 = 0.13$
    Turn $= 1.84/3.35 = 0.55$

2. (a) The bar chart (Fig. A.3) shows ornithine decar-
         boxylase (1), cytochrome c (2), aldolase (3), tyro-
         sine aminotransferase (4) and RNA polymerase (5).

    (b) A bar chart was chosen since the data describe
         discrete data sets. Notice that three of the five
         points cannot be distinguished therefore consider
         using two graphs.

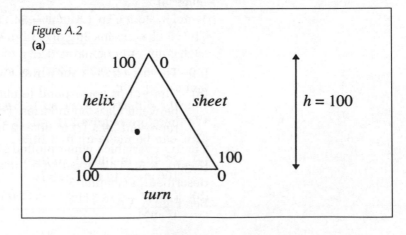

Figure A.2
(a)

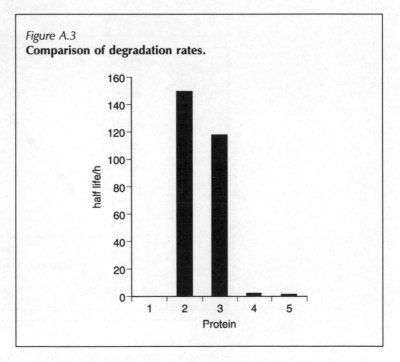

Figure A.2
(b) Preference of lysine for three structures.

Figure A.2
(c) Preference of three amino acids for secondary structures.

Figure A.3
Comparison of degradation rates.

3. (a) The bar chart is shown in Figure A.4.

    (b) The groups correspond to classes and show data between class boundaries. For example, 1 would represent 39.5 (?) to 60.5, 2 is 60.5 to 70.5 etc.

4. The two variables when plotted give a straight line (Figure A.5). This is called a linear relationship and is described in Chapter 7.

5. See Figure A.6.

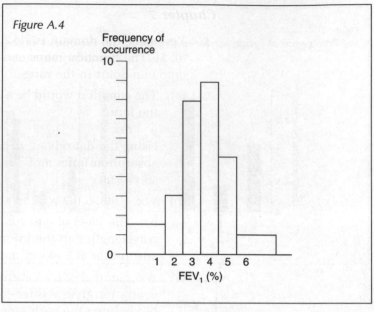

Figure A.4

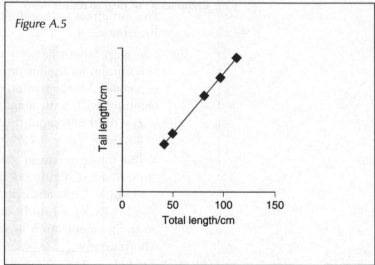

Figure A.5

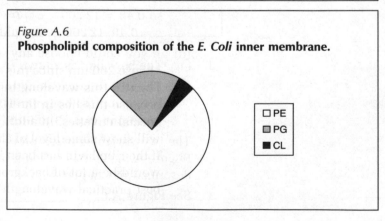

Figure A.6
Phospholipid composition of the *E. Coli* inner membrane.

Chapter 7

1. Yes, since the domain is {1, 2, 3, 4, 5} and the range is {0, 1}. The function maps each member of the domain onto one point in the range.

2. (a) The equation would be a special linear equation of the form:
   $y = mx$
   Using the data given, $m = 0.45 \div 0.005 = 90$ absorption units $mol^{-1} litre^{-1}$
   so $y = 90x$

   (b) $y/x = 90$ so $0.2/x = 90$ so $x = 0.2/90 = 2.2$ mM

   (c) No. The level of absorption is dependent on the wavelength and the information given is for adsorption at 540 nm, not 500 nm.

3. (a) $A \propto c$ and $A \propto l$, so a three-fold increase in concentration gives a three-fold increase in absorption but halving the path length halves the absorption. The net effect is therefore a $1\frac{1}{2}$-fold increase in absorption: $A = 0.72$.

   (b) The path length is twice that used in the initial experiment so the absorption is twice that expected. An absorption of 0.06 unit would be obtained with path length 1 cm. An absorption of 0.06 would correspond to a concentration of:
   $(0.06/0.48) \times 6 = 0.75\% \,(w.v)$

   (c) A 1 M solution would have an absorption equal to the molar extinction coefficient so:
   1 mol litre^{-1} has absorption 12 200
   The 2% (w/v) solutioin has absorption 0.48
   So the concentration is: $0.48/12\,200 = 38.3\,\mu M$
   Alternatively,
   $A = \varepsilon c l$
   so $0.48 = 12\,200 \times c \times l$
   $c = 0.48/12\,200 = 38.3\,\mu M$

   (d) If you could choose any wavelength, I would choose 260 nm since this absorbs most strongly. In practice this wavelength may lead to problems because this lies in the UV range so it requires special cuvettes. In addtion, DNA and proteins will show some level of absorption in this range so if the riboflavin has been isolated from a cell there would be a lot of background in this region. The best practical wavelength may well be 450 nm.

4. For a straight line the gradient is constant. If any of the three points are used the gradient is found to be 1/2. Since $\Delta y/\Delta x = 1/2$ in all cases, these lie on a line of which the equation is:
    $y = \frac{1}{2}x + 1.5$

5. (a) The Lineweaver–Burk equation has the form $y = mx + c$ where $y = 1/v$; $x = 1/[S]$; $m = K_m/V_{max}$; and $c = 1/V_{max}$. The symbol v represents the velocity (nmol litre^{-1} min^{-1}), [S] substrate concentration (M), V_{max} maximum velocity (nmol litre^{-1} min^{-1}), and K_m the Michaelis constant. From Table 7.6.
    Let $x_0 = 1/8 \times 10^{-6} = 1.25 \times 10^5 \, M^{-1}$ and
    $x_1 = 1/1 \times 10^{-5} = 1.00 \times 10^5 \, M^{-1}$
    Then (from Table 7.6) $y_0 = 1/13.8 \times 10^{-9}$
    $= 7.3 \times 10^7$ (nmol litre^{-1} min^{-1})$^{-1}$
    and $y_1 = 1/17.0 \times 10^{-9}$
    $= 5.9 \times 10^7$ (nmol litre^{-1} min^{-1})$^{-1}$
    $m = \Delta y/\Delta x = 560$
    so, substituting for x_0, y_0 and m into $y = mx + c$,
    $c = 3 \times 10^6$ (nmol litre^{-1} min^{-1})$^{-1}$
    so $V_{max} = 1/c = 3.3 \times 10^{-7} \, M \, min^{-1}$ or 33 nM min^{-1}.
    $m = K_m/V_{max}$ so $K_m = 560 \times (3.3 \times 10^{-7})$
    $= 1.9 \times 10^{-4} = 0.19 \, mM$

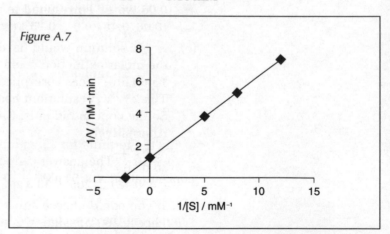

Figure A.7

(b) Figure A.7 shows the Lineweaver–Burk plot for the data. The values for 1/[S] have been multiplied by 10^{-4} for ease of plotting and are measured in M. The $1/V$ values have been multiplied by 10^{-7} and are measured in M min^{-1}. V_{max} can be read off the y intercept and gives V_{max} as 80 nM min^{-1}. $-1/K_m$ can be read off the x intercept and gives K_m as 0.04 mM.

(c) The plot is more accurate beause the line of best fit is obtained from more than two data points so that any error is averaged out.

Chapter 8

1. (a) $3x^2 + 4x + 2$ gives a discriminant less than zero so this equation has no real roots

   (b) Using $a = 2$, $b = -5$ and $c = 3$ the discriminant implies two real and distinct roots. The equation gives $x = 1.5$ or 1.

   (c) Using $a = 2$, $b = 7$ and $c = 3$ the discriminant implies two real and distinct roots. The equation gives $x = -0.5$ or -3.

2. (a) The volume increases by $5^3 = 125\%$

   (b) Surface area increases by $5^2 = 25\%$

   (c) The bacterium transports nutrients across the membrane and since the volume increases at a greater rate than the surface area the point will arrive where there can be no further expansion because the cell cannot take up nutirients fast enough to ensure its survival, i.e. you would predict that growth will be limited by a surface:volume ratio.

3. Rearranging the equation gives
   $$V_{max} = (v(K' + [S]^n)) \div [S]^n$$
   $$V_{max} = (5 \times 10^{-6}) \times ((25 \times 10^{-6}) + (1 \times 10^{-3})^2)$$
   $$\div (1 \times 10^{-3})^2 = 0.005 \, \text{M} \, \text{min}^{-1}$$
   so $V_{max} = 5 \, \text{mmol litre}^{-1} \, \text{min}^{-1}$

4. $K_w = [H^+][OH^-]$
   so $10^{-14} = (x + 10^{-6}) \times x$
   $$x^2 + 10^{-6}x - 10^{-14} = 0$$
   Using the formula for a quadratic, $x = 5.1 \times 10^{-5} \, \text{M}$ or -5.1×10^{-7}. The answer cannot be negative so the concentration $x = 51 \, \mu\text{M}$

5. Comparing the expected and observed number of each genotype we get

Genotype	Observed occurrence	Expected occurrence
AA	58	59.3
Aa	37	37
aa	5	5.8

It seems likely that the population is at equilibrium.

Chapter 9.

1. $10\,\mathrm{h} = 600\,\mathrm{min}$ so this is $600/40 = 15$ doubling times
   $N(15) = 10^3 \times 2^{15} = 3.38 \times 10^6$ cells

2. (a) Let the half-life $= \tau$
   $N(\tau) = N_0/2$
   so $N(\tau)/N_0 = 0.5 = e^{-\lambda\tau}$
   $\ln 0.5 = -\lambda\tau$ so $\tau = -\ln 0.5/\lambda = 0.69/\lambda$

   (b) $\lambda = 0.69/14.3 = 0.048$

   (c) (i) $N(t) = 6735 \times e^{-(0.048\times 4)}$
   $= 6735 \times 0.824 = 5552\,\mathrm{Bq}\,\mu\mathrm{M}^{-1}$

   (ii) The specific activity is $5552\,\mathrm{Bq}\,\mu\mathrm{M}^{-1}$ so $2367\,\mathrm{Bq}$ corresponds to
   $2367/5552\,\mu\mathrm{M} = 0.43\,\mu\mathrm{M}$

3. From Chapter 9 we know growth can be modelled using the equation:

   $$G(t) = G_0(1+y)^n$$

   Here $n = 10$, $G(t) = 360\,000$ and $y = 0.04$

   so $\qquad 360\,000 = G_0(1+0.04)^{10}$
   $$= G_0 \times 1.48$$

   so $G_0 = 360\,000 \div 1.48 = 243\,203$

4. Growth rate is $0.2/3.2 = 0.0625$ or 6.25% per month
   $W(t) = 3.2(1+0.0625)^4$
   $= 4.08\,\mathrm{kg}$

5. (a) The value after one month is the geometric mean of the value at zero and two months.
   Geometric mean $= (42 \times 48)^{1/2}$
   $= 44.9\,\mathrm{cm}$

   (b) $H(t) = 42(1+0.14)^6$
   $= 93.6\,\mathrm{cm}$

6. A geometric series would produce the best results. The ratio between the first two measurements is 1.5. This would be constant for a geometric sequence so the next five values are:

   $7.5 \times 1.5 = 11.25\,\mu\mathrm{M}$ $\quad$ $11.25 \times 1.5 = 16.87\,\mu\mathrm{M}$
   $16.87 \times 1.5 = 25.31\,\mu\mathrm{M}$ $\quad$ $25.31 \times 1.5 = 37.97\,\mu\mathrm{M}$
   $37.97 \times 1.5 = 56.95\,\mu\mathrm{M}$

7. Let $C_0 = 1$ unit, then a decrease of 5.6% gives
   $C_p = 0.944$
   $C_p/C_o = e^{-Kt}$
   $\ln 0.944 = -K \times 60$

$0.057/60 = K = 0.001 \, \text{min}^{-1}$

So 0.1% is eliminated each minute.

Chapter 10

1. (a) $\log x = 5 \log y = \log(y^5 x)$

   (b) $2 \log t - 4 \log t = \log t^2/t^4 = \log t^{-2}$

   (c) $0.5 \log((9m)^2) = \log((9m)^2)^{0.5} = \log(9m)$

   (d) $\log(a+b) + \log(a-b) = \log((a+b)(a-b))$
   $= \log(a^2 - b^2)$

2. (a) $\log 5x = 3.7$ so $5x = 10^{3.7}$
   $$x = 10^{3.7}/5 = 1002.4$$

   (b) $\log(4m - 3) = 0.9$ so $4m - 3 = 10^{0.9}$
   $$m = 2.74$$

   (c) $\ln x = 1.8$ so $x = e^{1.8} = 6.05$

   (d) $\log 2x + 3 \log x = 2.2$
   $$\log(2x \times x^3) = 2.2$$
   $$2x^4 = 10^{2.2}$$
   $$x = (158 \div 2)^{1/4} = 2.98$$

3. (a) $10 - 7.4 = 3.9 \times 10^{-8}$ or $39 \, \text{nM}$

   (b) (i) $6.3 \times 10^{-8} \, \text{M}$ implies that there is twice the concentration of hydrogen ions; therefore the blood is more acidic than it should be.

   (ii) $\text{pH} = -\log(6.3 \times 10^8) = 7.2$

   (c) Using the equation for pH the $\text{pH} = 7.2$

4. $\log K = \frac{1}{2.3} \times -E_a/RT + \log A$ $(\ln x = 2.3 \log x)$
   so $\log K = -E_a/2.3R \times 1/T + \log A$

5. The Hill plot gives a straight line with gradient 2 so there are two binding sites.

6. $\text{pH} = pK_a + \log([B]/\{BH^+\})$

   (a) Substituting the values for pH and pK_a:
   $[B]/[BH^+] = 0.5$ so $[B] : [BH^+] = 1 : 2$

   (b) At pH 2 the ratio changes to $[B] : [BH^+] = 1 : 5 \times 10^5$

   (c) The uncharged form (B) is the active form; therefore in the gut at acidic pH there is very little of the drug that is active.

7. A plot of $\ln C_p$ against t gives a straight line with the gradient $-K$ and y intercept of C_0.

Chapter 11

1. (a) Mode = 3.7 kg
   Median = 3.7 kg
   Mean = 3.49 kg

   (b) 95% confidence level using mean $\pm$(SEM $\times$ 1.96)
   3.49 $\pm$ 0.28 kg or 3.77 to 3.21 kg

2.

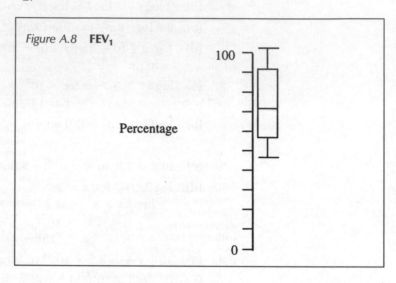

Figure A.8 FEV$_1$

   (b) Mean = 74%
   95% confidence using t test with 9 degrees of
   freedom gives:
   mean $\pm$(SEM $\times$ 2.262) = 74 $\pm$ (5.47 $\times$ 2.262)
   95% confidence interval is:
   74 $\pm$ 12.4 or 86.4 or 61.6%

3. (a) Peptide 1 mean = 11.82 mN m^{-2}
   Peptide 2 mean = 2.38 mN m^{-2}

   (b) Using the t distribution with 3 degrees of freedom
   for peptide 1 the 95% confidence interval is given
   by:
   mean $\pm$ (SEM $\times$ 3.183) i.e. 11.82 $\pm$ (0.35 $\times$ 3.183)
   95% confidence interval is 11.82 $\pm$ 1.11 or 12.93 or
   10.71 mN m^{-2}.
   Using the t distribution with 5 degrees of freedom
   for peptide 2 the 95% confidence interval is given
   by:
   mean $\pm$ (SEM $\times$ 2.571) i.e. 2.38 $\pm$ (0.13 $\times$ 2.571)
   95% confidence interval is 2.38 $\pm$ 0.33 or 2.71 to
   2.05 mN m^{-2}.

Index